William Halsted Wiley, Sara King Wiley

The Yosemite, Alaska, and the Yellowstone

William Halsted Wiley, Sara King Wiley

The Yosemite, Alaska, and the Yellowstone

ISBN/EAN: 9783744797788

Printed in Europe, USA, Canada, Australia, Japan

Cover: Foto ©berggeist007 / pixelio.de

More available books at **www.hansebooks.com**

THE

YOSEMITE, ALASKA,

AND THE

YELLOWSTONE.

BY

WILLIAM H. WILEY

AND

SARA KING WILEY.

Reprinted from "Engineering."

LONDON :

OFFICES OF "ENGINEERING," 35 & 36, BEDFORD STREET, STRAND, W.C.

NEW YORK :

JOHN WILEY & SONS, 53, EAST TENTH STREET.

PREFACE.

——•——

I HAVE been asked to introduce this book and its authors to the reader. To the American public such an introduction is not only superfluous, it is almost impertinent; to the British public it is unnecessary, because when the English reader has perused the book he will be on more friendly terms with the authors than any words of mine can make him. He will have become acquainted with some of their principal characteristics—their energy, power of observation, and facility of description; their honest whole-souled patriotism and pride in being citizens of the United States.

To undertake a pleasure trip of 10,000 miles is no light task; to carry it through successfully in the company of 30 or 40 fellow travellers speaks volumes for the feelings of friendship, harmony and self-control that animated the band of latter-day pilgrims from first to last. And to do all that in about two months, in comfort and without any mishap except the little untoward incidents that add zest to travelling, bears the highest possible testimony to the present condition of railway organization in the United States, and to the ability and organising power of Messrs. Whitcomb and Co., the Cook and Son of America. Such a journey could not fail to impress the most phlegmatic American (if such a being exist) with feelings of pride at the magnitude and grandeur, the wealth and the resources of his country. And when Americans are enthusiastic and observant, as both our authors are, what wonder is there that they should be almost appalled at every stage of the journey, most of which had, up to that time, been to them but a vague dream.

The circumstances under which this trip was carried out were peculiar and thoroughly American. It was neither more nor less than a part of the

summer convention of the American Institute of Mechanical Engineers.
This body, although it is one of the youngest of the technical societies of
the United States, has grown so rapidly that its members claim for it to-day
the foremost place amongst similar institutions.   Certainly it does not lack
in wealth of members, in excellence of organization, or in energy.   A part
of its business is to hold periodical meetings in other cities than New York,
where its head quarters are.   Prior to last year it had been far afield, and
had met in many places quite remote from New York.   But the idea of
convening at San Francisco, 3000 miles distant, seemed too bold and
difficult of accomplishment to any but a few of the hardier spirits of the
Institution.   Still these prevailed, and the result was the decision of the
Members to hold their summer meeting on the Pacific Coast.   Of course
these meetings are accompanied by visits to various points of interest,
and it was evident that the excursions on this occasion must harmonize in
scale with the business journey.   Naturally, therefore, one of them was the
visit to Alaska, involving the trifling journey of two or three thousand
miles, while others of less magnitude were confined to the glowing lands of
the Pacific slopes.   Our authors were of the Alaska group, and I know of
no record of American travel written by Americans, so fresh and bright
and vivid as that contained in the accompanying pages which I have had
the pleasure of editing (a merely nominal duty) and seeing through the
press.   It has partly repaid me for the disappointment I felt on being compelled to decline the pressing invitation I had received, to join the party.
Although I have had much experience, I have been unable as yet to
measure the length and breadth and depth of American hospitality, and I
resented bitterly the pressing duties at home that prevented me from
making one more effort to gage those dimensions.   I knew, however, when
I learned of my good friends' intention to publish a record of their journey,
that I should find much compensation in my disappointment, and the result
has justified my confidence.

I would say to all those who have never made this journey but are
intending to do so, that they can not do better than take this little book
along with them to serve them as a guide ; while those, less fortunate, may
gain, with a little effort, some good idea of what nature has done throughout
that vast stretch of country, and also what the energy of man has been
able to accomplish there during the last fifty years.   For the possible
benefit of both, I may complete this preface with a few dry facts and
statistics which would have been out of place in the pages of my friends'

story, but which may be interesting for reference to whose who desire to gain an idea of the vastness and power of the United States and of its sixty-six millions of people. I suppose that most persons imagine that the City of San Francisco marks the western boundary of the great Republic. So far from this being the case, it is practically the geographical centre. Reckoning from the meridian of New York, San Francisco is between 46 and 47 deg. west, while the western boundary of Alaska extends beyond 90 deg. Thus the United States of North America stretch in an unbroken line—save for the incursion of the State of British Columbia—one quarter round the globe.

All the vast area constituting together the western States of the Union is being rapidly occupied and its natural resources developed, thus adding year by year at an ever-increasing rate to the riches of the United States. A few words may be added about a few of these great States touched upon in lighter vein by our authors, who could well be excused, in the hurry of travel and under the constant influence of new emotions, from dwelling upon the more serious industrial aspect of the country through which they passed.

Taking first the State of Colorado—the *Silver State* ; apostrophised thus by Joaquin Miller, " Colorado, rare Colorado ! Yonder she rests ; her head of gold pillowed on the Rocky Mountains, her feet in the brown grass, the boundless plains for a playground ; she is set on a hill before the world, and the air is very clear so that all may see her well." So recently as 1840 Colorado belonged to Mexico, and its vast area of over a hundred thousand square miles was added to the United States in 1846. Fourteen years later it had a population of 34,000 ; it was transformed from a territory to one of the States of the Union in 1876, and at the last census, its population had increased to 410,000. Agriculture and mining on a prodigious scale are the main industries of the State, although various manufactures to no mean extent are also carried on. The climate is not well adapted for agriculture on account of extreme dryness, early and late frosts, and the plague of locusts, but the farmers more or less overcame these difficulties, and now over three million acres are in cultivation. To compensate for want of rain a system of irrigation has been carried out by which the water is brought from the mountains and distributed through more than 35,000 miles of canals and ditches. The value of the farm products is not less than three million sterling a year, and, in many parts of the States, fruit, including peaches and grapes, is grown in large quantities. Alfafa

a

clover is, however, the leading product, and three million tons of this
crop were grown in 1889. Stock-raising, and especially sheep-farming, is
also another important industry. The flocks of Colorado, according to the
last returns, numbered two million head, and from these ten million pounds
of wool are sent yearly to the eastern states. Mining industries, chiefly of
gold and silver, add largely to the riches of the State and of the Union.
They was inaugurated in 1858 near Denver, now a city of about 110,000
inhabitants. Altogether more than sixty millions sterling in gold and silver
have been extracted; to this must be added £10,000 worth of lead and
over a million sterling worth of copper, produced entirely from gold and
silver bearing ores. Almost unlimited iron deposits, chiefly hematite and
magnetite, with from fifty to sixty per cent of metal, are stored up for
future generations, for at present this mineral is but little worked. The
coalfields cover an area of 40,000 square miles, and the thickness of the
seams averages 5 ft. There are only 50 coal mines in operation in
Colorado, but they employ 5,400 men, who now raise annually 2 millions
and a half tons of coal. Petroleum, too, exists, and the twenty-five wells
that are in operation produce annually 150,000 barrels for lighting, and
about the same amount for lubricating purposes.

Montana, the *Bonanza* State, was traversed 150 years ago by a group
of French-Canadian explorers, and sixty years later the first settlements
were made by a Missouri fur traveller. It was not, however, until 1829
that the State was even nominally settled, and at the present time its
population is but 132,000, or about .3 to the square mile. It is essentially
a region of mountains and rivers, of rolling plains and dense forests; the
total area is 146,000 square miles, and these include 30 millions of acres of
farm land, 38 millions of grazing land, and 26 millions of woods and
mountains. The Missouri river takes its rise in this state and runs through
it for a distance of 1300 miles; the Yellowstone also traverses Montana for
a distance of 850 miles, after rising in the Yellowstone National Park in
the adjoining State of Wyoming, the beauties of which are so graphically
described in the following pages. Montana was about the last State to
shelter the now practically extinct North American buffalo; only about
twenty years ago vast herds of them, estimated at about 8,000,000, found
shelter in the 'western' plains, but fourteen years later these had been
practically exterminated. The progress of Agriculture in Montana has not
yet reached a high state of development, and indeed the difficulties of
climate are very great, nevertheless the head of cattle number one million

and a half, and the flocks yield annually as much wool as those of the State of Colorado ; horse-breeding is also an important and profitable occupation, and the State contributes about one million oxen annually for export to the East and to Europe. Mining, however, is the chief industry of Montana, and about 80 million sterling of gold and silver have been produced ; then lead and copper are also found in great quantities. The coal measures cover an area of 60,000 square miles, and the seams vary in thickness from 6 to 20 ft. ; 300,000 tons are raised annually.

California, or *Eldorado*, was discovered in 1634 by Spanish travellers ; Sir Francis Drake followed not long after, and regardless of any prior claims he christened the country New Albion ; but Spain maintained her footing on the Pacific Coast and the history of the Jesuit Missions for nearly 200 years, is one of the most interesting chapters in the records of North America. In 1822 the Colony renounced Spanish authority and allied itself with Mexico, and for many years the only foreign trade that existed was monopolised by Boston, whose merchant fleet made two-year voyages along the Atlantic and Pacific shores of North and South America. In 1846 the famous Captain Fremont and a small band of followers entered California overland ; they were speedily attacked and defeated by the Mexicans, but gathering strength by the accession of stray Americans, they made head against the Established Government and founded the California Republic. This was the beginning of the end of Mexican rule, and in 1848 the State was ceded to the National Government and admitted to the Union in 1850. At the present time its population is about one million and a quarter, and its area is 158,000 square miles. It was in 1849 that the discovery of gold took place, and the dreamy Pacific State was invaded by armies of fortune seekers who brought with them confusion and lawlessness to which the country became a prey for several years. All that, however, is a thing of the past, and now the State of California is one of the chief glories of the Union, as it is one of the richest. For many years mining was the ·chief source of wealth, but now agriculture is taking the foremost place. Between the years 1849 and 1890 gold and silver to the value of 250 millions sterling were produced. Until recently the system most largely used was that of hydraulic mining, and to furnish the water required, 5000 miles of aqueduct were constructed. So much damage was done, however, to the agricultural interests by the vast floods thus poured into the rivers, that hydraulic mining has been stopped by law except in one or two ocalities. The other minerals produced by California are quicksilver (two million pounds a year); copper, iron, salt, borax, sulphur, soda, antimony and coal. California ranks next to the State of Pennsylvania in its yield of petroleum, and large stores of natural gas are known to exist, although they are not yet developed. It is, however, in agriculture that the State of California is now most famous, for the climatic conditions are favourable for almost every class of farming or fruit growing. The cereal, hay, and root crops of the State are valued at fifteen million sterling yearly, beet sugar factories flourish, and at least one in operation is able to deal with 500 tons of roots daily. In the cultivation of fruit California is the foremost State in the Union ; it is estimated that there are about 25,000,000 of fruit trees under cultivation, comprising the whole range of the products in temperate and in some tropical regions. The orange, lemon, and olive flourish with the peach, vine, and apple. About 2,000,000 boxes of rasins are sent east yearly; in 1890, 33,000,000 pounds of grapes were exported, as well as 2000 tons of dried peaches, 4000 tons of dried prunes, and the produce of 300,000 fig trees. It was in 1858 that the State was regarded as suitable for vineyards, and at the present time 200,000 acres are planted with vines which produce more than 300,000 tons of grapes yearly, from which about 20,000,000 gallons of wine and 1,000,000 gallons of brandy are produced. The vineyards of California are 600 miles in length and 100 miles in width. The largest separate vineyard covers 4000 acres, and one of the wine cellars is on a corresponding scale, for it has a capacity of two millions and a half of gallons. This industry, however, was not a new creation, for it had been introduced by the Jesuits 200 years before, and the grapes grown, and the wine made by them, on a very small scale of course, have not yet been surpassed. Probably, when more experience has been gained, and time has been given somewhat to exhaust the too generous soil, the world may have to look to California as its chief source of supply for wine. The State possesses 6 million sheep, and the average production of wool is 35 million of pounds yearly. The cattle number about one million head, and the stock yards, though not to be compared with those of Chicago, are never-theless on an enormous scale. Efforts are being made to develop California into a great silk producing State, and the advantages of climate promise to give satisfactory results. Of smaller agricultural industries it may be mentioned that the bee was acclimatised in 1854, and now in two counties alone there are 50,000 hives, from which 6 million pounds of honey are obtained yearly. Ostrich farming has also been introduced on a comparative large scale, not with any marked result however.

Washington, the *Evergreen State*, one of the latest admitted to the Union, and the most remote from New York, has made prodigious development from the date of its settlement as a territory, in 1845, to the present time. Setting aside the more or less mythical stories of its discovery in the sixteenth century, the credit belongs to the enterprise of Boston merchantmen who, shortly after the Revolution, sailed through those remote and unknown seas that form its western boundary. From that time till 1845 the only inhabitants besides native tribes were the agents of the fur companies—English and American—and the ownership of the region was disputed by the Stars and Stripes and the Union Jack. But in the year just named, the United States took possession, and defined the boundaries of the territory, which gave it a length of 230 miles and a width of 350 miles. Although situated in a high latitude, the climate is comparatively equable and mild, and the country is well adapted for agricultural and pastoral pursuits. About 20 million acres are timber lands, 10 million are arable. 5 million of pastoral lands are credited to the State, and there are 10 million of wooded mountain. The wheat crop of Washington is at least 15 million of bushels annually; it is claimed to be the finest hop growing country in the world, and large quantities of fruit are produced. The tobacco crop is also becoming of great importance. As for hops, the present output is 40 or 50 thousand bales a year, and large quantities are brought to England. The lumber trade is, however, the most important in Washington, and the production in 1890 was no less than 1,201 millions of feet. The famous Douglas pines are found in this State; they reach a height of 300 ft., give spars often 150 ft. long, and planks 90 ft. in length. On Puget Sound is the largest saw mill in the world. It turns out 90 million ft. of timber a year; uses 3000-horse power (water and steam); and employs 250 men. Mining will be in the future one of the great industries of Washington, and is already of no mean proportions. Coal is raised to the extent of 1½ million tons a year; gold mines are profitably worked, and the iron deposits are apparently boundless. The Great Western Iron and Steel Company, with a capital of a million sterling, has commenced operations at Kirkland on Lake Washington. Of the chief towns in this State, vivid descriptions are given in the following pages.

Alaska, one of the few remaining Territories of the Union, was owned by Russia until 1867, when it was purchased by America for 7,200,000 dols. Its population is decreasing apparently, for in 1880 it was 33,000, and at the last census only 30,000. Except for the fisheries, it produces but little at present, though the indications are pretty clear that in the future it will be one of the richest, if not the richest portion of the Union. At Juneau, placer mining is carried on with profit, more than a million dollars having been obtained, and about the same total value is exported yearly to San Francisco. The Whalers from San Francisco and further east, obtain £300,000 a year in oil and bone and ivory ; there are 30 million cans of salmon and 15,000 barrels exported annually. The cod fishery is on a very large scale, and of the value of the sealing trades, something is known in this country, on account of the complications that have arisen between our-selves and the United States, but which doubtless, will soon find a satis-factory and friendly solution.

The actual condition and the potentiality of the Western States of North America might be enlarged upon indefinitely, but I have said enough perhaps by way of introduction to our authors' story. Those who seek further information will find it in that admirable _vade mecum_ King's Handbook of the United States, to which I place myself under a further obligation by reproducing the following table showing the value of the natural productions of the United States in 1890. The figures tell their own story.

STATISTICS FURNISHED BY THE DEPARTMENTS OF WASHINGTON.

		$
Corn	... ...	598,000,000
Wheat	... ...	342,000,000
Oats	... ...	172,000,000
Barley	... ...	27,000,000
Rye	... ...	11,000,000
Buckwheat	...	7,000,000
Total Cereals		... $1,157,000,000
Meats	... ...	740,000,000
Poultry Products	...	190,000,000
Butter and Cheese	...	245,000,000
Milk Consumed	...	160,000,000
Wool	... ...	66,000,000
Hides, Hair, &c.	...	93,000,000
Total Animal Products	...	... $1,494,000,000
Cotton	... ...	293,000,000
Market Gardens	...	70,000,000
Orchard Products	...	160,000,000
Other Products	...	655,000,000
Total Miscellaneous	...	... $1,178,000,000
Aggregate Farm Products		... $3,829,000,000

				§
Iron	...	...	...	107,000,000
Silver	...	...	...	60,000,000
Copper	...	...	...	34,000,000
Gold	...	...	...	33,000,000
Lead	...	...	...	16,000,000
Coal	...	...	...	212,000,000
Building Stone		...	...	55,000,000
Natural Gas	...	...	...	23,003,000
Petroleum		...	...	18,000,000
Salt	...	...	...	5,000,000
Other Minerals		...	...	21,000,000

Total Mineral Products... $584,000,000

Grand Yearly Aggregate $4,413,000,000

Without any misgiving on my friends' account as to the reception which "The Yosemite, Alaska, and the Yellowstone" will obtain at the hands of the English speaking public on both sides of the Atlantic, I offer them in advance my congratulations, on the artistic character of their volume, and on their fresh and novel treatment of a noble subject.

JAMES DREDGE.

London, March, 1893

LIST OF ILLUSTRATIONS.

— ⋅•⋅ —

THE YOSEMITE, ALASKA, AND THE YELLOWSTONE.

WHILE the journey, of which a description will be attempted in the
following pages, is intimately associated with the famous trip of the
American Society of Mechanical Engineers to their San Francisco Convention in the spring of 1892, yet there are mingled with this special experience
so many matters of considerable interest that the writers trust the narrative
will be found sufficiently attractive, simply as the story of the journey,
to warrant its perusal without the introduction of any detailed reference
to the real business of the Convention.

The starting point of the mechanical engineers was the office of
Messrs. Raymond and Whitcomb, of New York, whose "vacation"
excursions are the delight of every one who has ever participated in them.
It is not the object of the writers to "boom" these gentlemen, much as they
may deserve it, but the members of the society would consider their
historians as ungrateful in the extreme if they failed to chronicle how much
they were indebted to the careful and thoughtful plans of this firm, not
alone for their comfort, but absolutely for their happiness. To begin with,
we furnished simply ourselves to be taken care of, lodged, fed, transported,
and entertained. We had not even the care of our hand-baggage, while as
for our trunks, they were in a baggage car, and could be opened at any
time during the journey. It was most emphatically the luxury of
travelling. The train was one of the best equipped in the railway service,
and consisted of special sleeping-cars with numerous state-rooms for family
parties, and a dining-car, while at the end was what is known as a combination car for lounging, smoking, and reading; it contained a bath-room,
barber's chair, and other accessories of modern civilisation. The train was
drawn by one of the best engines in the West Shore service. The speed
was frequently a mile in 55 seconds, but that never interfered with either
reading or writing, for the West Shore line was built in the most thorough

A

B

manner possible, and at so small a regard to expense as to leave the original subscribers to wonder how so much money could have been spent in so short a distance. To the engineer, however, there is little difficulty in seeing that the road was constructed to stay, and that every engineering obstacle was met by a free expenditure of money.

A railroad map always is instructive, and the reader is recommended to study the one given of the West Shore Railway in order to better understand the commencement of the trip (see Fig. 1 annexed). The first duty of the maker of such a map is to show his company's line, even if only a small one, very straight, and in heavy black, and to make all others, especially all competing lines, as crooked as possible, and as lines in the strictly geometrical sense, with plenty of length, but with neither breadth nor thickness. Length indeed is the great desideratum; they must be longer than the route shown, and if unfortunately they are not, why they have to be made so, even if the face of the country has to be changed. The West Shore map is no exception. The course of its line is so broad as to blot out the New York Central, even where the latter is a four-track road, and Lake Ontario is conveniently pulled down at its western corner to show a short route to Hamilton. Of course, when the New York Central map maker starts in, he reciprocates in kind, and so "honours are easy."

Starting from New York early in the morning of May 4, we looked our last at the city as we crossed the Weehawken Ferry, gazing over the beautiful harbour, its shores lined with high-masted steamers, its waters alive with boats of all sizes, with the statue "Liberty," grey in the distance, and the gilded dome of the "World" building shining out over the irregular roofs of the city. At Weehawken we entered our special train and sped out across the Jersey meadows, the birthplace but, unfortunately, not the abiding place, of the world-famed mosquito. As we flew along, the gaunt lines of telegraph poles seemed to stalk away across the brown stretches of plain to the low hills on the horizon. For a short distance our way lay through fertile New Jersey, just springing into green life, with the farmers ploughing in the fields; then we passed into New York, and as we issued from a tunnel near Nyack, a view of the Hudson burst upon us, an expanse like the sea, bounded by lines of green hills, Storm King opposite shelving to the water's edge, and an outlook down the sweeping curves of the river, where the mountains seemed to open to let it pass and to stand aside one behind the other. This is aptly called the Gate of the Hudson, and here in earlier times stretched that iron chain which

Fig. 1.

FIG 1. MAP OF THE WEST SHORE RAILROAD AND CONNECTIONS.

Washington placed to keep back the British cruisers. We now passed through a tunnel, and under the parade-ground at West Point, where our Government has probably the finest military academy in the world. an institution on which it did not call in vain in that great struggle for the preservation of the Union in 1861.   Fig. 2 shows the road at this point, and in the centre above, may be seen the roof of the Riding Academy, where the cadets are taught to stick to a horse till the force of gravity insists on its rights being recognised, and then to fall gracefully into the sawdust.   In the distance appeared the Hotel Kaaterskill, perched on the top

FIG. 2.   TUNNEL UNDER WEST POINT.

of the mountains and overlooking the valley.   Those of the party who had attended the Convention of the Civil Engineers, held there some years ago, found their minds running back to that pleasant and festive occasion, and recalled the picture shown in Fig. 3, which is an engraving of the Catskill Falls in this neighbourhood.   We now began to slip along the Mohawk Valley, and parallel with that great artery of trade, the Erie Canal.   On the opposite side appeared the trans-continental train of the New York Central, intending to show us how easily it would throw dust in our eyes. But our engine was quite equal to its task, and its engineer had no wish for a second place ; so after running five miles in 4 min. 50 sec., we left the Blue Train to its own devices.   There were none of the usual discomforts of travel on this trip.   The cars were well ventilated and easy running, and even the element of dust was eradicated, for the country was brilliant from recent rains, the brooks were brown and turbulent, and pools of water were in the fields.   It was now time to visit the combination car, to smoke and

FIG. 3.   THE CATSKILL FALLS.

lounge, and, as Washington Irving has said, "he who smokes thinks like a philosopher," so there were those who went to the writing-rooms and chronicled their thoughts in imperishable ink, or, selecting a book from the shelves of a well-chosen library, found thus one of the pleasant pastimes of the journey. The sight of the Erie Canal brought to our minds the apparent indifference of our legislators to this great highway ; for while our more or less hostile Canadian neighbours have been deepening and enlarging the Welland Canal, so that a man-of-war could easily pass through, we do little more than scrape away the mud periodically in our canal, although its enlargement and deepening would be neither difficult nor expensive. At about nine we reached Niagara and going into the vestibule as we passed across the bridge, we looked down on one side and saw the dark tranquil river gliding between black lines of trees, while from the other side we heard the roar and crash of seething water and saw the white flash of the foaming rapids. This scene is familiar to many, and the reader is referred to Figs. 4, 5, 6, 7, and 8, for some very characteristic views of the falls, and of the rapids above and below them.

FIG. 4. BRIDGE BETWEEN SISTER ISLANDS, NIAGARA.

On Thursday morning we awoke to hear the clatter of a driving storm and to see the level green landscape without, show dimly through a wavering curtain of rain. A little later we found ourselves in a very flood—fields and villages inundated. One man, wiser than his fellows, had

FIG. 5. THE HORSESHOE FALL

FIG. 6. THE ROCK OF AGES.

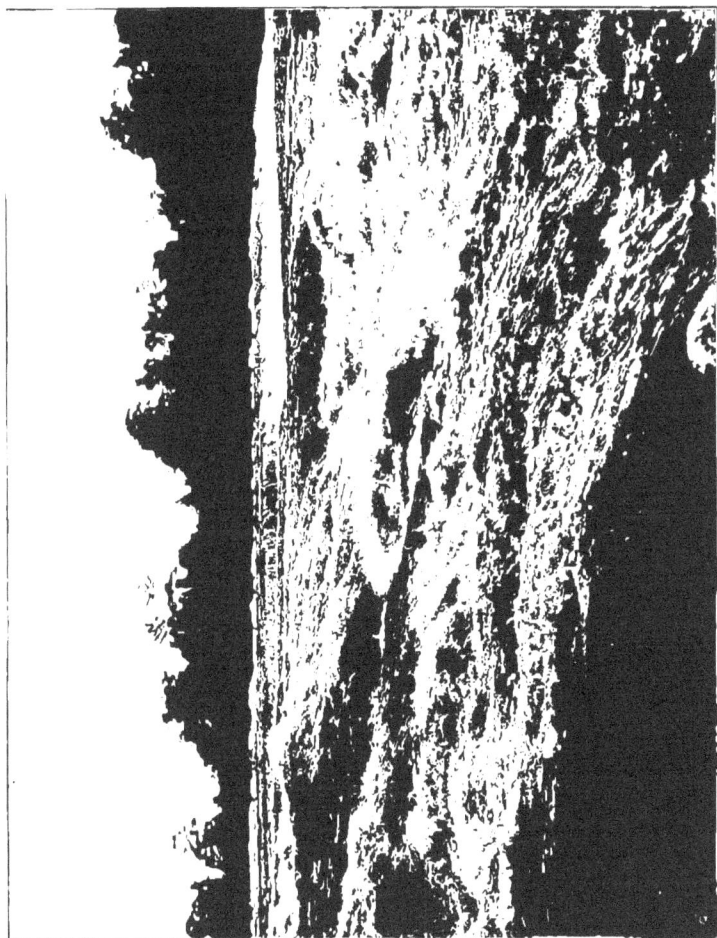

FIG. 7. THE RAPIDS ABOVE THE FALLS.

Fig. 8. The Rapids below the Falls.

c

a boat tied to his fence, and in another yard the bedraggled chickens were assembled in a mournful group on the wood pile, apparently reflecting as to when the Ark and Capt. Noah might be expected.

One of the great attractions on the Chicago and Grand Trunk line, which we next entered upon, and which was advertised as a cogent reason for taking it, was the "wonderful St. Clair Tunnel, that great triumph of engineering skill," that "subjection of the forces of nature," &c. Whether it was not on exhibition just then, or was out of humour, or what not, it is impossible to say, but we passed through during the hours of the night ; however, the writers were bound to see it, and although it can hardly be said truthfully that they really saw it, they were wide awake enough to make affidavit that they heard it, and that we rushed through it at a good fair speed. Nothing of note occurred on this part of the trip, and in due time, after a most excellent breakfast provided for us by our conductors, and quite superior to the best Pullman service we had ever experienced, we reached Chicago about noon, that is, we reached its limits, but in fact we stopped about five miles outside—you can stop a long way outside Chicago —and from here we took carriages and drove to the Exposition grounds. These grounds were somewhat out of repair, and were a complete refutation to the statement we have no ruins in America. Here we first rode in the ingenious travelling walk and saw the Women's Building, which was sufficiently constructed to give a clear idea of its appearance. It is to be a handsome and artistic structure ; the noble and graceful figures sustaining the roof being particularly noticeable. The enormous skeleton of the Building of Mechanical and Liberal Arts next claimed our attention. The facts about this structure are simply overwhelming. It is the largest building in the world—everything, by the way, in Chicago is the "largest in the world ;" they can't help it, it so happens—being five times the size of the Coliseum at Rome, and having an area of over 30 acres. The Administration Building will be a very handsome structure and the colossal statues which are to ornament it were examined with great interest. All are strong and beautiful, but the group "Patriotism" is especially fine. The sight of rows and rows of enormous heads, wings, and limbs on the floor was novel and interesting, but vaguely suggestive of a nightmare. Do not let the reader think our tour of the Exposition ground was made in carriages ; so extensive are they that we were obliged to go by railroad, and even then had but little time at each point. The mechanical features were the ones which were most attractive to this party, and as the chief engineer and some of his assistants were members of the Society and were present, we had rare opportunities for seeing what is to be, as well as to comment on what was.

Re-entering our carriages we next drove towards the heart of the city down Drexel Boulevard and Madison Avenue, admiring the many broad roads stretching in all directions and the lines of handsome residences. After visiting the machine works of Fraser and Chalmers we returned to the train, where we learned that "washouts" had blocked all exit from Chicago, and that we must remain overnight. By the courtesy of Mr. Wm. Chalmers, who first sent beautiful bouquets to our train for the ladies and then invited the entire party to attend the theatre, we all passed a delightful evening, and next morning walked about the city until eleven, when we started westward, proceeding slowly through the flooded country. In many places the tracks had been only temporarily sustained. The great tracts of level pasture land looked like a shining lake; in one place we could see ten miles of water, with here and there a small house appearing submerged to the second floor. As the afternoon drew on the scene was a beautiful one; the gold and crimson of the sunset clouds dyed the waste of waters, and the delicate new foliage of the trees was pale yellow against the deep blue sky. In some places the train crawled slowly over a track that waved in curves frequently reversed, while the rails being shored up on temporary cribwork, trestling, and the like, caused the car to tip from side to side, which kept up a certain amount of interest in our progress. A curved line is undoubtedly a line of beauty, but a vertical curve on a railroad track does not conduce to æsthetical thoughts. The water in many cases covered the rails entirely, and photographs were taken at frequent intervals, for we had a full complement of the Kodak fiend, in fact, they were so largely in the majority that, perhaps, they should be considered as in the normal condition, while we, who were not, might be called the anti-camera lunatics. One of the most attractive bits of scenery on the Rock Island Route is given in Fig. 9, and a glance at the map will show the locality as being between Chicago and Rock Island. The mapmaker has again kindly taken out all curvatures, and the reader will notice that the Rock Island route is positively a bee-line from Chicago to Omaha. If he don't find it so in practice that is no fault of the mapmaker, who certainly has done his best in the case.

That evening we reached Rock Island, where the Government has one of its finest arsenals, and crossed the Mississippi under the soft light of the full moon. The river was greatly swollen by the recent rains, and looked almost like the ocean, but the tracks were so high that the floods caused no delay nor even anxiety.

Saturday morning, May 7, found us speeding over the fertile pasture lands rolling in green billows to the horizon and dotted with herds of cattle, the horses in their winter coats looking as if they were sadly in need of a shave or were developing into the celebrated winged steeds of antiquity. At Council Bluffs we stopped a short time, and one of the party bought a

FIG. 9.   HORSESHOE CAÑON, OTTAWA, ILLINOIS.

paper, which banner of the hustling West was dated Sunday, May 8 ; an issue twenty-four hours in advance seeming to be considered the proper sort of enterprise in a western journal. However, he was equal to the occasion, and said immediately, "Here, I don't want this back number, give me Monday's paper."

The cameras had been diligently at work all day, and the owners met to compare notes, when it appeared that one gentleman who had been particularly active and who was under the impression that he must have nearly exhausted his films, found that he had been taking pictures all day on a white paper roll, which is used to teach beginners, he having, in his desire to get everything, entirely forgotten to remove this practice sheet. During this day's trip we were frequently in doubt as to our further progress, for we were crossing the level country in Nebraska, and the water was on all sides. To add to the complication the engine seemed tired (of course the wheels were so), and had a fashion of getting off the track to rest at intervals, and this involved considerable hard work and much harder language on the part of the train men. After two of these rests it appeared that a tender axle was bent so as to compel the greatest caution in proceeding, and just

FIG. 10.   THE CITY OF MANITOU.

then it began to rain as if it never had rained before. The storm increased and the waters were rising, again threatening to cut off further progress, but fortunately for us we obtained another engine and sped off westward at a rapid pace. The danger was not imaginary, for the waters rose so rapidly as to cut off travel for a week, our train being the last one which passed. That night the entire party assembled in one car and had a sort of variety show. There were orations, historical stories, banjo playing, and musical performances. The thunder furnished the applause and the lightning illuminated the darkness outside, while the tremendous rain on the roof was a suitable accompaniment. All this went on as we sped west into Colorado at 60 miles an hour. The next morning we waked up into a bitterly cold climate, and over the bare brown stretches of prairie drove a whirling snowstorm, followed by clouds of grey fog. A tiny little calf separated from a herd of cattle stood shivering in the wind and roused our sympathy as we shot past him. We reached Manitou, shown in Fig. 10, at 9 A.M., and in front of us were the gaunt, ragged foot-hills of the Rockies, their low pines and greyish furze powered thick with snow ; while Pike's Peak formed a background of grandeur unsurpassed. Although it was not actually raining the sky was overcast and the air laden with moisture. Carriages were in readiness, and we drove to the wonderful Garden of the Gods. On every

FIG. 11.   TEAPOT ROCK, GARDEN OF THE GODS.

side were the strange red rocks in numberless fantastic forms as though tossed by Titans ; gigantic toadstools, weird and comic faces, animals, and even fish. Some of these are to be noted in Figs. 11, 12 and 13. Wild yellow sweet peas and unknown bell-like blossoms of blue and purple grew among the short sparse grass. Before us was the gateway to the garden rising 400 ft. in height, shown in Fig. 14, looking through which we saw the low hills, their delicate grey-green shades contrasting beautifully with the deep red of the great boulders. Passing through the gateway we turned and saw the Rockies, towering through drifting mists, peak behind peak, snow-crowned, vast, and rugged, with Pike's Peak capped in snow for a gigantic background. Then passing on through the garden the clouds to the westward swept away and the pure silver sky shone out behind the dark-green foothills, while above, the white mist coiled and uncoiled about the ragged summits, and a sudden rift in the clouds let a single ray of sunlight fall on the gleaming snows of Pike's Peak, which rises to some 14,400 ft. above the sea level. On we drove, past the clashing Rainbow Falls to the delicious iron spring and to the wonderful cave. Here, lamp in hand, we passed down a narrow pathway deep into the ground. Curious and beautiful stalactites hung dripping above us, with crystals glittering like diamonds.

FIG. 12. MOTHER GRUNDY, GARDEN OF THE GODS.

We stopped in one vast dim cavern, where the rocks over our heads shelved up into a great black void, and from a sort of balcony a man played on a natural organ formed of stalactites of varying size. The sweet but hollow music echoed weirdly through the great vault in faint tones. Then our guide struck another formation, which gave forth the deep ring of a chime of bells.

FIG. 13. CATHEDRAL SPIRES, GARDEN OF THE GODS.

FIG. 14. THE GATEWAY AND PIKE'S PEAK.

There are many interesting trips to be taken at Manitou, and if one could only be left to enjoy them at leisure and in peace, a much better impression would be given to the visitor. But you are continually called upon by the guide to admire this or that feature from the standpoint of some one else's imagination. For instance, while in the Garden of the Gods, where the mind naturally reverts to that time when this was undoubtedly the bottom of a great inland sea, whose waters have, by steady attrition for years and centuries, worn the rocks into the curious shapes presented, and when one thinks of the littleness of the present race and the short time of their tenure upon earth, it is most rasping to be suddenly startled by " Mister, that ere rock is called the Dutchman, since it looks like Hans, and next to it is Hans' wife and the baby."   Or, " This rock is the Eagle and the Bear, showing the eagle in the act of attacking the bear." You feel like attacking the speaker, and one of the ladies told him she had no imagination and never could see resemblances, and did not want to ; but this only stimulated him to try and make them so unmistakable that she must admit them. The visit to the Grand Cavern is worthy of the time and trouble, for, in the writers' opinion, it ranks next in beauty to the celebrated Luray Cave of Va. ; moreover, the trip itself through the Cañon (Ute Pass) and by Rainbird Falls repays the visitor. In the cavern is Grant's Monument made by piling loose stones in a pyramidal form, each visitor adding one.   Some bones are also shown you, said to be those of a cave dweller ; as they cannot be inspected, the chances are that the bones are much more modern and most probably those of some unfortunate cat or dog, who, having seen the Grand Cavern, was so overcome that it died there.   " See Rome and die," why not then do the same at Manitou ?   The writer preferred to live and to go to Cheyenne Cañon, not to see where " H. H." was buried, but to see the seven falls rising one above another to a height of 500 ft., as shown in Fig. 15.

All the various cañons are well worth exploration.   The scenery is so varied that there is no semblance of monotony, and each is impressively grand in its own individual way.   Nor should the tourist omit the ride to the summit of Pike's Peak.   In 1885, the ascent was extremely difficult, and had to be made on mule back, but now there is a railway, that a digression may profitably be made to describe.

The preliminary surveys for this road* were commenced in April, 1888,

---

* This description of the Manitou and Pike's Peak Railroad was contributed by Mr. J. G. True, of Denver.

under the immediate supervision of Mr. D. E. Briggs, chief engineer of the Denver and Rio Grande Railroad. On the 1st of April, Mr. T. F. Richardson, chief engineer, and his assistants, reached the top of the Peak with four day's rations, in a very severe snowstorm. It was impossible for them to leave the protecting walls of the old Government station, built of stone, until compelled to join the main body on account of hunger, and it

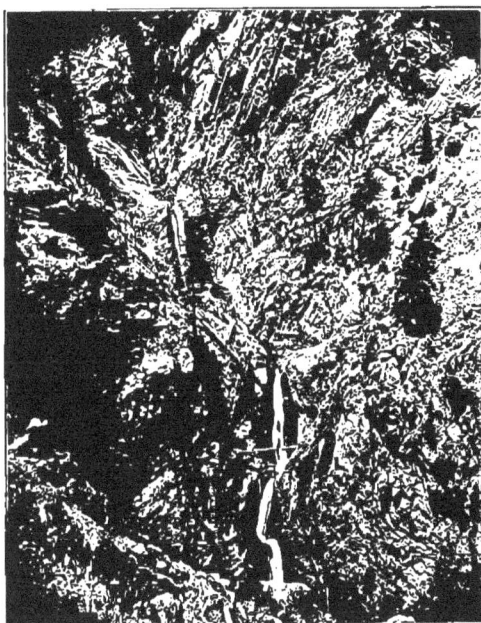

FIG. 15. THE SEVEN FALLS, CHEYENNE CAÑON, COLORADO.

took them ten hours to travel the one and a half miles to do so. Such experiences were frequent until June, when the snow had melted and the labour was less dangerous and more easy. On September 28, 1889, grading was begun at the summit, and before the winter storms set in, three miles of roadbed were completed. In 1890 three additional miles were graded, the road was practically finished, and some testing trials and excursions made.

The objective point, or upper terminus of the railway, which has been built for the same purpose as those up the Rigi, Mont Pilatus, and other summits of favourite summer resort, is at the top of the perpetually snow-mantled summit of Pike's Peak, 14,336 ft. above the sea level. Its lower terminal is Manitou. And here, before proceeding, let us say a word about Manitou. As with Pike's Peak—though its historic notoriety is less widespread—there are many thousands who either know or have been told of its wonders, and yet other thousands to whom the name sounds strange and meaningless. Lying securely protected among the picturesque foothills which cluster around Pike's Peak, and not more than seven miles, as a carrier pigeon would travel, from the giant's summit, is the most charming resort in the known world, and that is Manitou. The most nearly perfect climate yet discovered, the medicinal value of its numerous mineral springs, and the rarity and magnificence of its scenery, are the prime causes which give it claim to that distinction ; and since its first inhabitation, which dates but a few years back, its popularity and prosperity have been remarkable, and now its visitors number more than one hundred and twenty-five thousand annually, and are largely on the increase with each successive season.

Several years ago it was determined—contrary to the theory of Major Zebulon Pike, who discovered and named the promontory in 1806—that its ascent was possible, and the United States War Department, eager for so lofty a point of observation, established a signal station upon its highest point. With this the summit began to be visited by the more adventuresome tourists. Horse trails were constructed along the various watercourses, and travel over them increased rapidly. What is known as the Ruxton trail, bordering the beautiful Ruxton Creek directly from Manitou —the shortest and most interesting road (for the scenery approaching this mountain is indescribably grand)—was ever the most popular route, and was travelled by thousands every summer.

The railway has no counterpart on the American continent, though it is somewhat similar to the road at Mount Washington. It is of standard gauge, with wide and substantially built roadbed and heavy steel rails, the traction devolving upon two heavy serrated rails in the centre upon which operate six cog-wheels underneath the locomotive. It is built upon the Abt system (in use in Switzerland), and the peculiar mechanical construction of both track and locomotive render it absolutely safe. The length of the track is about $8\frac{3}{4}$ miles, or to be exact, 46,158 ft., in which there is a total ascent of 7500 ft. The steepest grades are 25 per cent., or a rise of 1 ft. in 4 ft., and there is very little of the track on a grade of less than $12\frac{1}{2}$ per cent. There are many curvatures, nearly 40 per cent. of the line being on curves, the sharpest of which is 16 deg. The bridges are entirely of iron and masonry, and the track in the steepest places is solidly anchored every 200 ft.

The present passenger equipment of the road consists of three locomotives and six passenger coaches. The locomotives are peculiar in appearance and weigh 25 tons each. They push the cars on the ascent and precede them on the descent, thus giving the engineer absolute control of the train should any breakage occur in the couplings. The coaches are elegant, largely of glass to facilitate observation along the route. Each has a capacity for fifty persons, and the seats are so arranged that passengers have at all times a level sitting. (See views in Figs. 16 and 17.)

The locomotives will each push two coaches and will make the trip up in less than two hours, including a stop at the Half-way House station, a beautiful retreat in Ruxton Park.

The care, safety, and rapidity of this new means of transit to Pike's Peak will add largely to the already large tourist travel in that direction. No visitor to Manitou, or even to Colorado, will wish to miss a trip so novel and beautiful, as will be seen from Fig. 18.

The following information has been supplied by Major John Hulbert, president of the company, originator of the scheme, and who has been untiring in his efforts to complete it. The entire length of the road will be 8¾ miles, and the roadbed is 15 ft. wide. Every 200 ft. or 400 ft., according to the grade, are sunk cross-sections of masonry, to which the track is tied, so that absolute rigidity is secured. There is not a single foot of trestle-work on the entire line, and only three short bridges, these being constructed of iron. The maximum curvature is 16 deg., which gives a radius of 359 ft. The average ascent per mile is 1320 ft. The total rise from base to summit is 7525 ft. The road is laid with 40 lb. steel rails ; between these, in the centre of the track, are placed two cog-rails, made of steel. This central rack rail is composed of two parallel steel bars 1¼ in. thick, and placed 1½ in. apart, in such a way that the tooth in one plate is opposite the space in the other plate. The pitch of the teeth is 4¾ in., and the depth 2 in. ; the rack is attached to every alternate sleeper, at distances of 3 ft. 6 in., by plate chairs, and shown in Fig. 20 on page 24, and fixed by ⅞ in. bolts. On the Mount Washington road, and on that up the Rigi, the middle rail is constructed upon the principle of a ladder. This is cumbersome, and only allows a speed of about three miles an hour. The speed attainable on the Pike's Peak road is intended to be 17 miles, but the

FIG. 16. VIEW OF THE TRAIN, MANITOU AND PIKE'S PEAK RAILROAD.

Fig. 17   Locomotive for the Manitou and Pike's Peak Railroad, Colorado

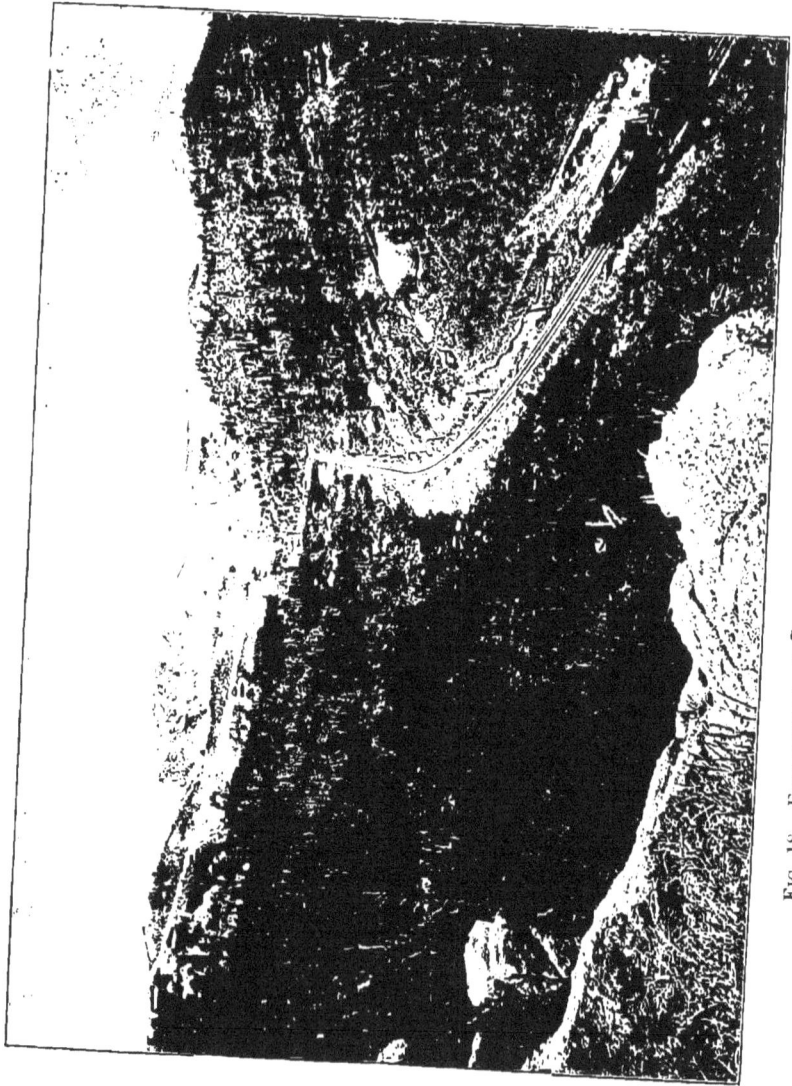

FIG. 18. ENGLEFIELD'S CAÑON, MANITOU AND PIKE'S PEAK RAILROAD

maximum rate will not be over eight miles, and the average not more than four miles. One cog-rail would be amply sufficient to do all the work, but two are inserted to insure safety. The engines are built by the Baldwin Company, of Philadelphia ; when on a level track, they stand at an 8 per cent. slant, and thus when the cars and engine are on the grade they are approximately level. The general design of the engine calls for no special remark ; the cylinders are 18 in. diameter, and the stroke 2 ft. ; to the bogie of the engine are attached the pinions that gear into the rack ; the arrange-

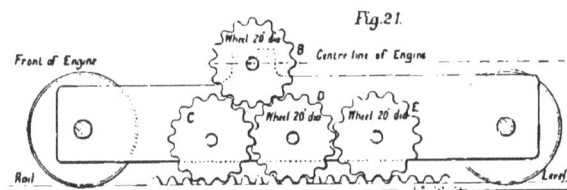

Fig.21.

Fig.18.

Fig.22.

Fig.20.

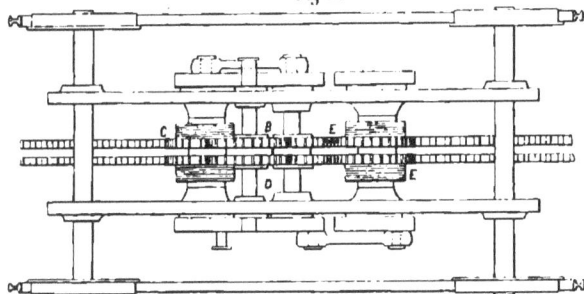

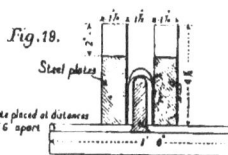

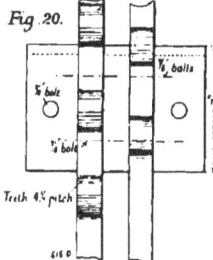

ment is shown in the diagram, Figs. 21 and 22. In the two vertical side frames of the bogie, through which the main axles pass, is mounted the crank-shaft B ; on it are keyed two toothed wheels gearing into C and D, the latter gearing into the wheel E, and C, D, and E engaging in the rack above described. A pair of spurwheels on each of the axles of the bogies of the car also gear into the rack, and brakes are mounted on every axle. There are three wheels on each side of the engine which revolve on the axles and merely act as guides, and to sustain the weight. There are three driving cog-wheels, which gear with the cog-rails. The weight of the engine is 32 tons. Two of the cog-drivers are in constant use, and the third is reserved for emergencies. The cars were built at Springfield, Mass., and are arranged on a "slant" corresponding with that of the engine, and each one is fitted with an independent cog-brake. The engines push the cars up the mountain, and are in front of them in making the descent. The cars seat fifty passengers each, though nearly twice that number can be accommodated in case of necessity.

As the ascent is made, many opportunities are given for exquisite views of the world below, through vistas in the trees, with the eastern plains glowing in the sunshine, and extending as far as vision reaches, and limited only by the blue horizon's verge. About half way up the mountain, and directly on the line of the railway, reached also by the trail, lies the Half-way House.

When the head waters of Ruxton Creek are reached, the road curves to the south-west, and "Windy Point" is attained. From here one has a distinct view of Manitou, Colorado City, and Colorado Springs. The "Cathedral Spires" and the "Great Gateway" of the Garden of the Gods appear like the castles set by the giants for a stupendous game of chess. We are now far above timber line. On all sides can be seen strange flowers, of lovely forms and varied hues. Plants which attain considerable proportions on the plains are here reduced to their lowest terms. It is not an unusual thing to find a sunflower stalk on the prairies rising to a height of from 8 ft. to 10 ft. ; here they grow like dandelions in the grass, yet retaining all their characteristics of form and colour. Beyond this mountain meadow, are great fields of disintegrated granite, broken cubes of pink rock, so vast in extent that they might well be the ruins of all the ancient cities in the world. Far below, flash the waters of Lake Morain, and beyond, to the southward, lie the Seven Lakes. Another turn of the track to the northward, and the shining rails stretch almost straight up what appears to be an inaccessible wall of precipitous granite. But passing the yawning abyss of the "Crater," the line proceeds direct to the summit. The grade here is one of 25 per cent., and timid passengers will not escape a thrill of fear as they gaze over the brink of this precipice, although the danger is absolutely nothing. At last the summit is reached, and, disembarking, the tourists can seek refreshments in the hotel, and then spend the time before the train returns, in enjoying the view and in rambling over the seventy acres of broken granite which form the summit.

E

The majesty of greatness and the mystery of minuteness are here brought face to face. The thoughtful mind is awed by the contemplation of this scene, and when the reflection comes that these great spaces are but grains of sand on an infinite shore of creation, and that there are worlds of beauty as vast and varied between the tiny flowers and the ultimate researches of the microscope, as those which exist on an ascending scale between the flowers and the great globe itself, the mind is overwhelmed with wonder and admiration.

In the directory of the Manitou and Pike's Peak Railway Company are many of Colorado's most prominent business men. Mr. D. H. Moffatt, president of the Denver and Rio Grande Railroad (the first narrow gauge railroad built in the Rocky Mountains); he is also one of Denver's most wealthy bankers. J. B. Wheeler, banker and mineowner of Aspen, Colorado; R. R. Cable and J. B. Glasser. Congressman Roswell P. Flower, of New York; Henry H. Porter, Chicago, president of the Eastern Illinois Railroad; Z. G. Simmons, of Kenosha, Wis., are among the heaviest stockholders.

The writer of this notice (Mr. J. G. True) desires to return thanks to Major John Hulbert, the president, and to Mr. R. S. Cable, general passenger and traffic agent, for their innumerable courtesies and attentions during his visit at their main station. He is also greatly indebted to Mr. J. G. Hiestand and his able assistant, Mr. L. E. Jones, for the views with which this notice is illustrated.

On returning from Pike's Peak, let the visitor stop at the Bath House and take one of the celebrated natural spring baths. The waters are charged with soda and iron, and the bather finds himself almost put to sleep by their soothing influence. A large swimming pool is attached, and affords a most delightful recreation. Having concluded the bath and the swim, the visitor is advised to take a large tumbler of Manitou ginger ale, well iced, and then if he does not say "life is worth the living" he had better die right there, for he will be useless anywhere else. After a good dinner at the Cliff House and the "Mansions"—for the party had to be divided—we started on our train for Colorado Springs, and from the hotel shown in Fig. 23, we took a farewell look at Pike's Peak, as shown there, and were soon speeding on our westward way. A glance at the Railway Company's map will show that the line from Colorado passes straight through the cañons, despite the fact that there are 15-deg. curves or more for very many points, no little matter of this kind weighs with a skilled map maker. We went but a short distance that night, for one of the greatest attractions of this entire trip was the arrangement made by Raymond and Whitcomb to have our trains put on the side track whenever desired, so that we could always see the finest scenery by daylight. We reached Cañon City quite early, and went up into the town, where the party received an invitation to attend a concert given by the Glee Club of the students of the School of Mines of Colorado, who were making a tour through the State to raise money for their library. After an enjoyable evening we returned to our train and went to bed just as comfortably as though in a hotel. Always having our dining-car, we were perfectly inde-

FIG. 23. PIKE'S PEAK AVENUE.

pendent of any hotels or restaurants, and had a uniform quality of food which could not be claimed for any eating places we saw unless on the plea it was uniformly bad in them. We saw plenty of restaurants, and many bore the sign, "Quick-Order House." One added, "A fine meal, 25 cents, and a perfect gorge 50 cents." As we expected the next day to have a Royal Gorge provided from Nature's own storehouse, this did not attract us, and sleep reigned profound in the car, only broken by certain sonorous notes from certain sections.

The train left Cañon City for the Grand Cañon at an early hour, and soon came to a halt at its entrance, as some supposed, to give us a view of the State Penitentiary, a fine stone building which contains 400 convicts, some of whom we saw working on the road under a guard; or, possibly, as some one suggested, to transfer some of our number to its fostering care, but after a census, in which each man denied having broken any law the previous night, although some admitted the lateness of their return to the train, the charge was disproved.

We passed Denver in the early hours of the morning, or late at night, according to the habits of the individual, and the writers think that they may be excused if they introduce at this place a description of the City of Denver and its surroundings, prepared by them some years since, on the occasion of the Convention of the American Society of Civil Engineers, held at Denver in 1886. Many changes and developments met them during their second visit, but on the whole they remarked little to induce them to modify their earlier impressions.

"To visitors from Europe who are accustomed to see a place grow slowly, and by gradual accretions, the apparently ephemeral growth of an American city in the far west seems phenomenal, and they fail to grasp the situation, yet here was a city full of life, bustle, and enterprise, with elegant stone buildings which will vie with any similar structures anywhere, all the growth of twenty-five years, and not a man who addressed us was a native of Colorado. Situated on the direct line of travel from the east to the west, and also on that from the north to the south, with all the products of California and of Mexico pouring into her borders, here at the base of the Rocky Mountains, Denver stands, once a city of the desert, but now environed by the arts of engineering, surrounded by a land that blossoms as the rose. Who can predict her future ? Her people have the energy of giants and the wise forethought of the Greek. Since twenty-five years the population is 70,000, and it has doubled within ten years. A great future is before her, and none were quicker to see this than the engineers.

"The first impressions of Denver are imposing. The railway station is a large fine stone building, and the streets leading from it are wide and attractive. Indeed, nothing is done here on a small scale. The population has increased rapidly to nearly 70,000 in a space of twenty-five years, the buildings are of a lofty and permanent character, mostly of the beautiful Colorado stone with its varied colours, while through the streets on either side flows a stream of clear water occasionally tapped so as to run around the base of a tree set between the sidewalk and the roadway. This at once attracted our attention to the secret of Colorado's success, and the irrigation system of which we had heard so much, was actually at our feet. As our hosts said, the engineers would at once, on their arrival, begin to ask questions, and as the climate is a dry one, it was thought advisable to lighten the labour of the Denver people as much as possible, for this is a great labour-saving country, so each visitor was presented with a book handsomely printed, and containing some 140 pages. Its title was significant, "Some Answers to Questions likely to be asked by the Members of the American Society of Civil Engineers during their Visit to Denver." It was extremely timely, and was seldom if ever consulted in vain. From it we learned that we were at an elevation of 5196 ft., but this was as nothing to our later experiences. The Continental divide is at no great distance from Denver, and the streams at its summit flow into the Atlantic or Pacific Ocean. We also found out that we were in a State that had shipped in minerals over 20,000,000 dols. in the last five years. We suspect the book of some irony when it states, "The air being dry the heat is much less felt than a lower temperature in damper climates." The writer has tried both, and confidently asserts that the damper climate acts as a shield from the intensity of solar heat. The water supply of Denver is on the Holly plan, and in several instances artesian wells have been sunk, and water of a beneficial mineral character has been obtained. If the water is alkaline in its character woe betide the party who drinks it, his throat will feel like the Desert of Sahara in a very few minutes, and the amount of fluid one can absorb in vain, is something unparalleled. Iced tea is a good beverage for Colorado, but this country is not one for prohibitionists; in fact, the old hymn applies here, "Ye thirsty souls fresh courage take." It is said to be a great country for invalids, especially for those troubled with throat or lung troubles, and it is so, for all the germs of disease must dry up and blow off. One railroad superintendent, who looks the picture of health, told us he had gone there originally to die; he had evidently changed his mind on that point. So much for Denver at the first glance; we soon learned to know it well, and to appreciate the great kindness and hospitality of its inhabitants, and having concluded our convention we started off the next day for Greeley to inspect the irrigation system. This is the glory of Colorado, and indeed without it the State would be almost a desert. In some parts of the State there is ample water supply, and this is fed by the snows of the Rocky Mountains, which are deep, and on many peaks remain all the year. By collecting this supply through canals and feeders, and distributing it over the portions needing it, the whole face of the country is rapidly changing. The present methods are rather crude, and are the immediate development of necessity; the gradients vary from 1 ft.

in a mile to 10 ft. ranging below 3 ft. The Citizens' Canal has a fall of
1 in. in 10,000 ft. at one point, while the Del Norte Canal, running through
a rock cut, falls 30 ft. in a mile.   The standard of measurement is a cubic
foot per second, and this it is stated will irrigate 50 to 55 acres.   The
annexed diagram shows the method of irrigation.   The view in Fig. 24
shows that portion of the canal running through Platte Cañon ; while
another illustration is given in Fig. 25 of one of the aqueducts.

"The entire canal system of Colorado embraces over 800 miles of large
size canals completed, and about 150 miles projected, and as well as 3500

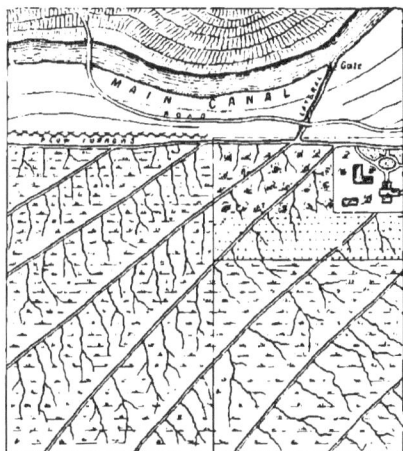

miles of canals of secondary size. The extent of the distributing territory is
about 40,000 square miles, and the entire system has cost from 10,000,000
dols. to 12,000,000 dols. The total area supplied is 2,200,000 acres, and
the arable land is 20,000,000 acres. In one case a flume crosses Bijou
Creek at a height of 30 ft. and is 2700 ft. long. These considerations
show what the enterprise of Colorado is, and how alive the people are to
the needs of their land. The Society was delighted with its investigations,
and prepared to enjoy the next day's trip, which was of a grand and
imposing character, being a ride to Georgetown through the beautiful
Clear Creek Cañon.

" Descriptions fail to present this ride adequately, and reliance must be
principally placed on the engravings accompanying this notice to convey to
our readers what we saw. The maximum grade is 211 ft. to the mile, and
the shortest radius is 478 ft. The first view is of Inspiration point, and is

FIG. 24. HIGH LINE CANAL, PLATTE CANON.

sufficiently inspiring to tempt any one to take the ride. It may be remarked here that the progress of civilisation has changed the waters of Clear Creek from their original colour. The creek is of a muddy yellow, due to the dirt washed in from the tailings at the upper end, but it rushes alongside the track with great force, and frequently leaps over a rock with a roaring sound at once melodious and startling.

"The engraving, Fig. 26, shows another sinuosity of a narrow gauge railway, and the engine plunged around this curve as though it was thoroughly used to it and really liked the fun.

FIG. 25. AQUEDUCT ON THE PLATTE CANAL.

"At Idaho Springs we found hot wells, swimming baths, and natives anxious to sell us 'specimens.'   These consisted mostly of bits of pyrites and galena ore, whose lustre renders them very saleable to strangers.

"At Golden there is a university and a mining school, and indeed there are the greatest facilities for a practical education in this branch. That it is sorely needed in Colorado is evidenced by the many unsuccessful prospectors whose abandoned claims point sadly to the passer-by the futility of human hopes, and seem to be warning ghosts of the past against rashness, and too great trustfulness.

"The view from Black Hawk from the passing train showed the smelting works and other industries, and suggested to the writer that there was more

FIG. 26. A CURVE IN CLEAR CREEK CAÑON.

F

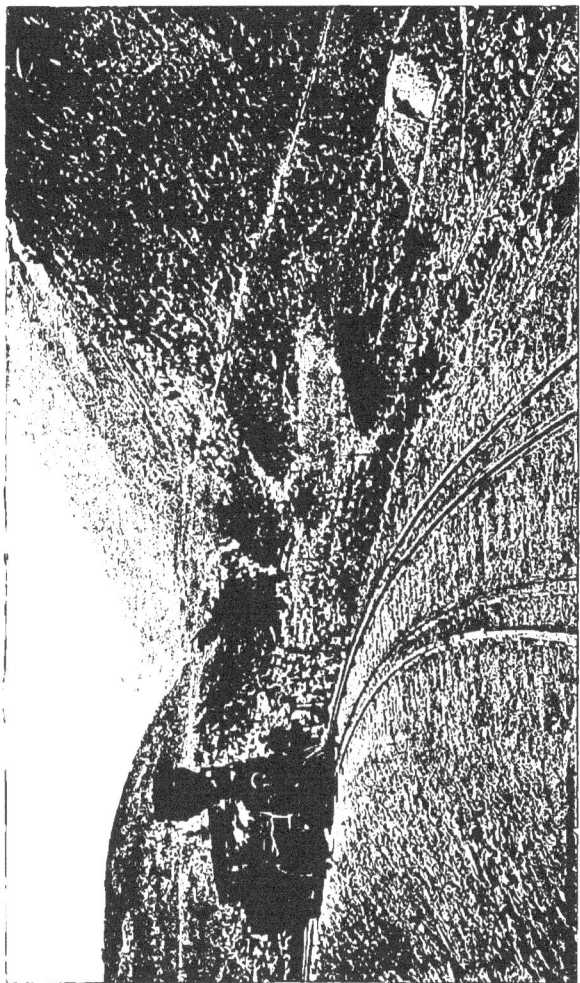

FIG. 27. UPPER END OF LOOP NEAR GEORGETOWN; UNION PACIFIC RAILWAY.

money in smelting than mining, just as the broker who does a commission business in stocks grows rich whilst his patrons lose millions.

" At last Georgetown came in sight, and we rapidly ran around the town and proceeded to the celebrated Loop, a plan of which is given below, and a careful study of which is necessary to properly understand the pictures given of this most peculiar and unique method of railway construction.

" The view Fig. 27 shows the upper end of the Loop, and those in Figs. 28 and 29, the general alignment of the railway, and the viaduct at the lower end It may be remarked that to make two miles, 3.9 miles are travelled, and an elevation of 623 ft. attained ; the average grade is 160 ft, to the mile, and the maximum grade is 190 ft.   At the point where the road runs under itself the viaduct (see Fig. 29) spans the valley ; the distance between the upper and lower tracks is 75 ft.

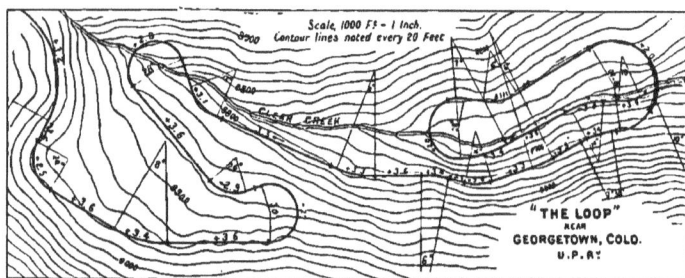

Scale, 1000 Ft = 1 Inch.
Contour lines noted every 20 Feet

CLEAR CREEK

"THE LOOP"
NEAR
GEORGETOWN, COLO.
U.P.R?

" Passing the Georgetown loop, and still ascending, we reached an elevation of nearly 9500 ft.; and arrived at Silver Plume, where the beautiful and now unpolluted waters of Clear Creek flow over a rock, and take a vertical drop, the water bursting into clouds of spray from which the name is derived.   This is quite an enterprising town, and is, as will be seen from the illustration, Fig. 30, on page 38, most romantically located.

" We had now reached the end of our day's ride in one direction, namely, upward, and the effects of the elevation had been marked to some extent ; we therefore waited for our engine to turn on a Y, and then returned to Georgetown for rest and dinner.   The writer there found an old friend who was the physician of the place, and on tendering him an invitation to come and dine, the doctor declined, because he expected a patient in the train shortly due, and they were so scarce he could not afford to miss one.   This speaks well for the salubrity of the place, and the expected patient had met with an accident.   People dry up and blow away.   It is due to the air, and

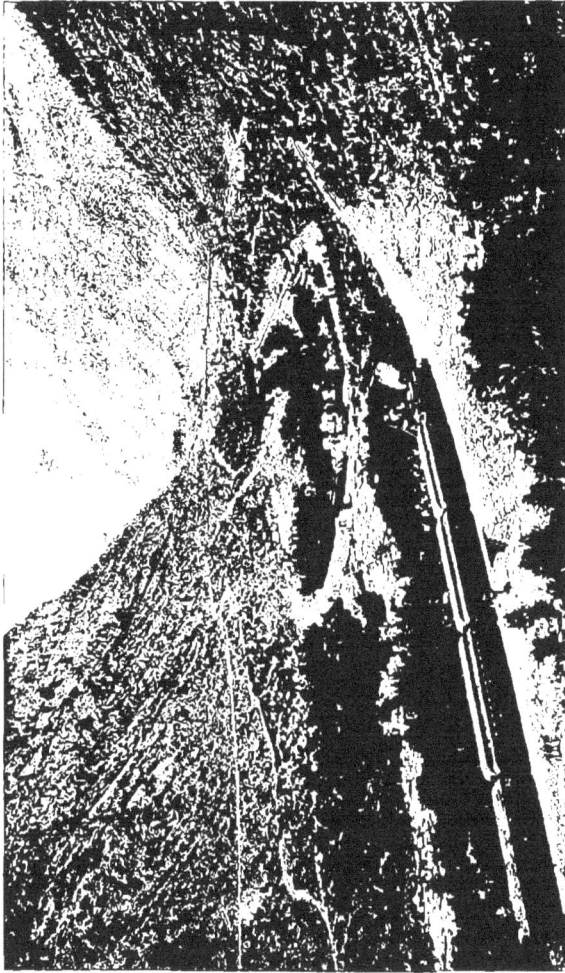

FIG. 28. LOOP NEAR GEORGETOWN; UNION PACIFIC RAILWAY.

FIG. 29. VIADUCT AT LOWER END OF LOOP, NEAR GEORGETOWN; UNION PACIFIC RAILWAY.

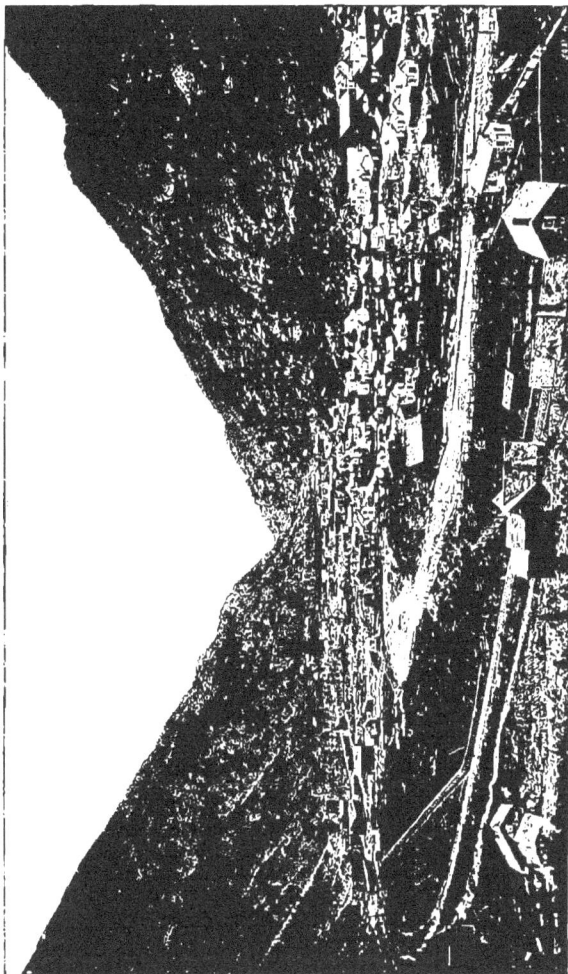

FIG. 30. THE CITY OF SILVER PLUME.

FIG. 31. PLATTE CAÑON.

is said to be healthy.   No oxen are seen here in the streets, and every one
uses horses or burros.   No cats are here, they have fits and die, owing again
to that 'air;' the inhabitants are not feline people.   It may be remarked
of this Clear Creek road, that in 13½ miles it falls 1700 ft., and it was
graded at a cost of 20,000 dols. per mile.

"Georgetown gave us a good dinner, and we returned to Denver to
rest and prepare for the trip to Leadville on the following day.

"The next day we bade farewell to Denver and started for Leadville,
the train having to be run in two sections, as the grades are so severe as to
limit the number of cars one engine can haul.   On entering the Platte
Cañon, one of the most beautiful in this wonderful country, the 'Dome
Rock' is one of the most striking and conspicuous objects.

"After following the creek at a rising grade and around curves of a very
small radius, similar to that shown in Fig. 31, till some of the party
became dizzy and others cross-eyed, we began to ascend in real earnest, and
soon found ourselves among the peaks of the Rockies, which seemed to rise
on all sides, and in every instance were covered with snow.   Along the
sides of the cañon were, as before, abandoned sluices, where many a life
had been worn out in the greedy search for the hidden treasure which
seemed to be ever almost within the grasp of the seeker, and which yet
managed to just escape his clutch.   The train stopped at the summit of
Kenostra Hill to let the visitors have a farewell view of these peaks, and
it is vividly impressed on the minds of all who saw it.   The summits in
the background stood boldly in relief against the clearest and bluest of
skies we had ever seen.   The pureness of the air and its rarefaction made
them seem very near, and the sparseness of the vegetation marked our near
approach to the 'timber line,' for we had attained an elevation of 10,139 ft.

"Shortly after crossing the summit we descended into South Park, a
most beautiful plateau extending some 65 miles, and being in breadth
about 45 miles.   A clear stream of water ran by the side of the track, and the
irrigating system was in vogue.   In the distance we saw herds of cattle,
and everything pointed to peace and plenty.   This latter idea was beginning
indeed to be a factor in the minds of many, and it was rapidly assuming
control.   On the distant hills could be seen a snowstorm raging, and we
entered Como amid a shower of hail.   The dinner here was all that could
be desired, and the town itself repaid a visit, although it was small, yet
there were plenty of signs of the usual Colorado enterprise.   And now we
started to climb to the highest elevation reached by a railway in this

FIG. 32. ROYAL GORGE, GRAND CAÑON OF ARKANSAS.

country. The railroad winds around the mountain side, circling towards
its summit, and the little town of Como becomes merely a speck below.
As there was a rapidly falling stream near the track, as water had been
abundant, and the gold promising (it always promises), we found hydraulic
mining had been largely practised, the method of working being that
familiar to every one. The force of the water from the nozzles used is very
great, and the bank melts away like snow. That it is a most wasteful
system of mining is beyond doubt; but who, in the early stages of searching
for gold, ever considers economy? Many abandoned claims greeted our
eyes, but the miners were still working at others, and as we reached the
higher elevations they seemed like ants moving around the hill, and finally
became entirely indistinguishable. Still we climbed, sometimes running
through a snow-shed, and occasionally passing snow banks until, in
Breckenridge Pass, we reached an elevation of 11,498 ft. On entering
Fremont Pass, named from the general, the train stopped to give the
party a view of the Mount of the Holy Cross, some 30 miles distant, and
whose summit bears in white the outlines of a huge cross. The road again
winds around the mountains until we were like Dr. Holmes's boy, whose
trousers were made the same on back and front, and who couldn't tell
whether he was going to school or coming home. At one moment we
seemed to approach Leadville, and at another to be going away from it.
The greatest altitude reached was 11,540 ft., and it was quite as well that
we did stop at this elevation, for some of the ladies were affected by
faintness, and in one or two cases became unconscious. Several of the
party had severe nose-bleeding, but in general the members were of
'seasoned timber' and used to getting high. We ran into Leadville at
dusk.

"This town at an elevation of 10,200 ft., started in 1859 under the name
of California Gulch, and in five years 5,000,000 dols. of gold dust was taken
out by washing. It then fell into decay, and in 1876 silver was discovered
in the hills Its population at once rose to 30,000, and enterprises were
started in every direction; the hills are honeycombed with openings, and
many a man has become rich from almost nothing, and so likewise many a
rich man has lost all; the latter, it may be said, are usually the eastern
capitalists, who invest from a distance. Every method of mining may be
seen—sluicing, hydraulic, and lode mining. Leadville also has an opera
house and a large hotel; it boasts of gas, electric lights, and all the modern
improvements, including first-class gambling houses, &c. The people have,

however, learned to move deliberately—as we soon found when trying to hasten—with the best of reasons. A run of ever so short a distance at this elevation makes the heart beat violently, and all the pictures of the fighting in Leadville, where persons are running to the *melée*, may be set down from this circumstance alone as base frauds. The household cat is also here wanting, and we slept the sleep of the just and shone to the perfect day, or something like it—one gets metaphors sadly mixed in Leadville. The next day, after visiting Marshall Pass and the Black Cañon, we passed through one of the greatest wonders of this wonderful country, to wit the Royal Gorge, where the sides rise vertically, it is said, for 3000 ft., and the cañon is so narrow that it almost seems as though the cars would scrape on the opposite side.

"It will be noted in the view Fig. 32 that one side of the bridge, which is a plate girder, is hung from a truss also shown, which spans the cañon from side to side, and in that sense it may be called a suspended bridge. This method is not known to have ever been practised elsewhere.

"The river here is the Arkansas, and we were informed that some miles below it literally disappears, sinking into the earth. This may be a 'fish story,' but we did not like to tell our informant this, because a want of confidence in Colorado sometimes begets a want of vitality in the doubter, and we preferred to live long enough to write this notice. At all events the river at this point gives no signs of its impending fate, but roars merrily and lustily through the gorge at a sharp descent, occasionally rising or rather falling, to the dignity of a cascade. The grades now become easy, and the train reached Pueblo, which is said to be a pleasant place, in the afternoon. The climate of Pueblo is remarkably equable even in cold weather, and it is a great health resort for invalids, especially if the trouble is with the lungs.

"The same is said of Colorado Springs, which we reached later; but one of the most charming spots in Colorado, and one which we most regretted to leave, was Manitou, most delightfully situated at the base of Pike's Peak, which rises for a background to the altitude of 14,147 ft., and whose top is clothed with everlasting snow. This town is set right among the mountains, although only at an elevation of 600 ft. itself.

"The air is cool and balmy, and the adjacent wonders and the excellent hotels render Manitou one of the finest watering places in the world. The mineral springs are effervescing, and consist, one of soda, one of sulphur and one of iron. This latter bursts from the earth at a temperature of ice water, and is thoroughly impregnated with iron; in holding it up to the light, bubbles rise as they do in a glass of champagne.   It is consoling and strengthening to all partakers, and the hotel where it is a speciality is one of the best at Manitou.

" The first thing to be explored is the ' Garden of the Gods,' situated near the town ; it is full of enormous boulders, which have by the action of the water been worn into fantastic shapes.   This locality was undoubtedly the source of an immense lake, and these boulders show plainly how the waves have worn their sides and have earned for them the titles which posterity has affixed.   There was something rather awe-inspiring as we gazed on these relics of an age far beyond the memory of man, and the reflection of what scenes and what prehistoric monsters these quaint and curious stones had witnessed was forced on the beholder (see Fig. 33).

" The grandest sight of all was the view of Pike's Peak through the gateway of the ' Garden of the Gods.'   This is admirably shown in the illustration (Fig. 34).

" These rocks, which are of sandstone, rise on either side over 400 ft., and the white rock in the foreground only a short distance from them is limestone.   On entering the gateway, rocks of the most fantastic appearance are seen ; nearly one hundred are to be selected, which have various names due to their fancied resemblances, the favourite names being the Sphinx, the Cathedral Spires (which look like those of Milan), the Kissing Camels, &c.

" At the approach from Manitou, a large balanced rock stands as a sentinel who would challenge any sacrilegious visitor.   It looks as though the slightest push would precipitate it down the steep incline it dominates, and yet it probably stood where it now rests when the earth was young, and may stand there till the final disruption of all things.

" If this strange spot where Nature seems to have sported with huge masses of stone looks attractive by day, its weird forms look even more uncanny at night, and the silvery light of the moon enhances their strangeness and magnifies their mysterious influence.   As this was the proper season, every one had the moonlight view, and all were impressed with the grandeur of this most picturesque spectacle.

" The party for Pike's Peak started early, and made the ascent and descent in twelve hours.   The signal station at Pike's Peak is a most dreary place, and no one who is detailed there is an object of envy, as may be judged from the illustration (Fig. 35).

" The next day a party started for the Cheyenne Cañon, and it may be

FIG. 33. GROUP OF ROCKS IN THE GARDEN OF THE GODS.

FIG. 34. PIKE'S PEAK; ROCKY MOUNTAINS.

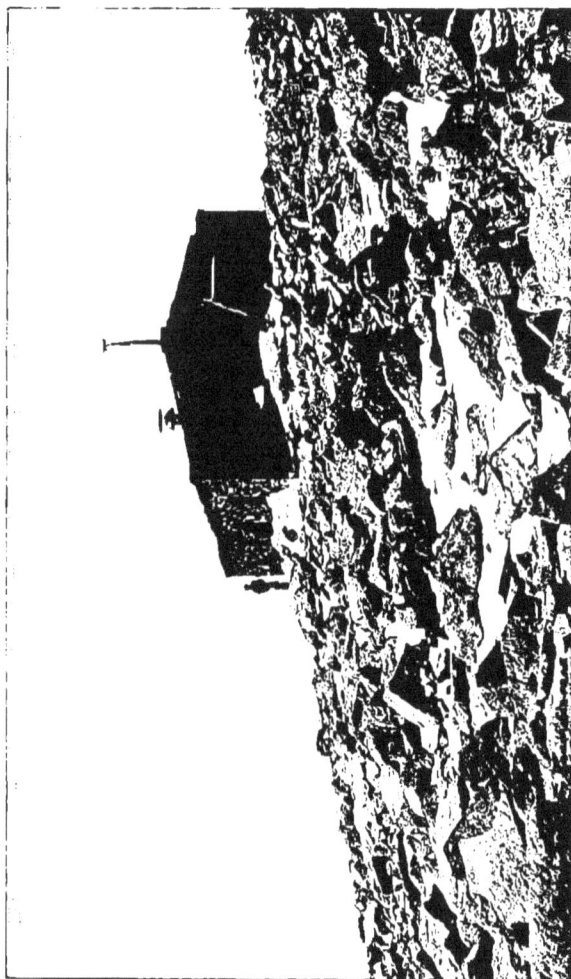

FIG. 35. U. S. SIGNAL STATION ON PIKE'S PEAK.

said that the millionaires of Manitou will be its livery stable men.   Not that their charges are so exorbitant, but their arrangements are so complete for preventing the traveller from seeing more than one thing in a day.   For instance, Cheyenne Cañon is by way of Colorado Springs sixteen miles distant, but why one must go by the springs was a question promptly asked of our driver.   He said there was no other way, although a road seemed to go in the direction of the Cañon from the 'mesa' or tableland.   After we had gone by the way of the springs, which occupied the day till 3 P.M., and was very much like going to Ludgate Hill from Charing Cross by way of High Holborn, our driver discovered a short cut for the return, which landed us in Manitou in 1¾ hours.   As the team was hired by the day, one can apply the reasoning, since we were told we must take it all day as it could not be done in less time. Nevertheless, in full view of all these facts, the writer unhesitatingly says, it paid.

" At the summit of this cliff is the grave of ' H. H.,' who has so abused our beneficent Government for its dealings with the Indians, and has thereby, in the writer's view, simply shown herself to be an extremely silly and impracticable woman.   However, he is a sceptic on her merits and has yet to see one of her books which in his judgment will repay perusal, her last, ' Romona,' being more impossible than any other.

"On the day following, we went to Ute Pass and saw the Rainbow Falls , a sketch of Ute Pass is given in Fig. 36.

" The falls are surprisingly beautiful, and the horses no doubt thought the road up the pass, of which an illustration is given, did not warrant any display of enthusiasm.   It was a fearfully steep pull, but at the summit was the celebrated ' Cave.'

" The writer is pretty well 'up' in caves (certainly we all were on this at the elevation).   Mammoth and Luray Caves have both been visited by him ; Luray Cave stands without a peer for beauty, but the Manitou cave is certainly remarkably fine, and well worth the climb it costs and the one-dollar entrance fee.   The stalactites are semi-translucent, of great beauty, and of marvellous shapes.   The ' organ ' is composed of a series of them alongside of each other, and when struck each has a different note, the result being that a skilful player can perform on them with great effect.   There are also many other beautiful and wonderful stalactites and stalagmites.

" On leaving Ute Pass, we rode to William's Cañon, of which the engraving in Fig. 37 gives an excellent idea.

FIG. 36. UTE PASS.

FIG. 37. WILLIAM'S CAÑON.

" In some places this was so narrow that the whippletrees scraped the walls on either side. The cliffs towered up above us in great magnificence and to dizzy heights, while the action of the river which had undoubtedly flowed through it at some time, had cut the rocks into the most fantastic shapes and castellated appearance. This too has its subterranean cave, but we took their word for it, and they did not take our dollar, so we returned to our hotel richer in pocket and with a new supply of faith. One thing more completed the Manitou trip to our great regret, and that was the swimming bath. This structure is very picturesque and very conveniently arranged. You can take a most luxurious hot soda bath and then plunge into the swimming tank some 200 ft. by 100 ft. and about 5½ ft. deep. When you have finished, and taken a good drink of the soda spring, you will feel so invigorated that you will unhesitatingly say " life is worth the living."

Resuming our journey, we entered the Royal Gorge, and were soon whirling along the cañon by the side of a most beautiful stream which continually disputed the right of way with the railroad track until the latter, apparently in sheer desperation, surrendered, and gave the insistent stream the whole cañon, where the track is suspended, and the stream passes under it. This chasm has been fitly named the Royal Gorge, and the height of the walls on either side may be set at nearly half a mile. To the spectators gazing from the rear platform up at the narrow band of blue sky, the cliffs seem to plunge headlong in wild and scattered confusion to the swirling foam of the stream at their foot; looking backward they tower one behind another, the sunlight slipping down here and there through a deep gorge and lighting the mica, the red sandstone, and the green serpentine on their sides, while the smoke of the engine rolls back in the clefts. At the end of the gorge a single enormous boulder stands like a sentinel, and on all sides the pale green hills sweep away in broad curves. Here, in the sand between the tracks, grew delicate white anemones. The irrigation trough followed the railroad for a long distance through this cañon, and is one of many such encountered in Colorado. By the aid of these great engineering auxiliaries the waste places have literally been made to blossom as the rose.

Suddenly, through a turn in the road between the side hills, burst a view of the snow-crowned Rockies, white against the bright blue sky, their summits wreathed with masses of grey clouds.

Soon we were up in the foot hills, on whose rocky soil only cactus grows, or dank scrub pine. In the plains we had just left were the houses of the prairie dogs, and we saw several of the cunning little beasts scudding away or cocked up in front of their holes with their tiny paws drooping in front of them. Now we were steadily climbing, and the air became chilled from the snow above us as we "cork-screwed" up the mountains in elaborate curves and loops, so that we could look back and see our former path far beneath us. More and more barren grew the vegetation, and near us were patches of snow. Soon we were beside great drifts and surrounded by white peaks. The air was fresh and invigorating, the sky full of drifting clouds. In the distance could be seen the peaks of the Sangre de Cristo, rising to 14,000 feet, and on the opposite side were the Arkansas Hills. The scenery, although beautiful, was somewhat subdued in appearance, and frequently meadow lands were to be seen, while herds of cattle lent a peaceful look to the landscape. We reached Salida about noon, and the entire party distributed themselves among the stores which contained the two desiderata of tourists, viz., souvenir spoons and photographs. After lunch, taking narrow-gauge cars, the party started for Marshall Pass, which is one of the grandest and most beautiful trips to be taken on this continent. The route runs through Poncha Pass, which acts somewhat like the prologue to the play, and serves to prepare the mind of the visitor for what is to follow. The route is circuitous, and crosses from side to side of the pass, always rising. The curves range upward till a maximum of 24 deg. is reached—it used to be 34 deg.—and the grade likewise gets more and more steep till 211 ft. to the mile is attained. The progress is very slow, and this gives a most delightful opportunity to observe the scenery. The road turns forward and back in a manner which kept the sun in a most uncertain state ; no sooner would it shine in the windows on the right side of the car, and evidently feel the satisfaction of warming the people, than a turn of the road would transfer its rays to the left, and, later, to the rear. The compass needle became discouraged, for north and south constantly changed places with east and west, but one thing was true to itself, and that was the ENGINEERING aneroid which was given to the writer many years ago. That instrument had made up its mind thoroughly what was its mission in life, and mounted higher and higher till it pointed with pride to an elevation of 11,000 ft. The scenery had grown bolder, and the vegetation proportionally sparser. Snow drifts and patches of snow appeared frequently, and the track so doubled on itself, that to gain an elevation of 1000 ft., we went five miles and faced in a direction directly opposite to our starting point. We were, in fact, like the celebrated boy of Dr. Oliver Wendell Holmes, whose mother made his pantaloons the same way in front and behind, and hence he never could tell whether he was going to school or coming home ; if the reader thinks the narrator is romancing let him carefully examine Fig. 38, where four lines of the same railroad are shown. Mount Ouray appears there in all its glory, and Mount Shavano is also to be seen ; both these mountains take their names from two celebrated chiefs of the Ute tribe. Ouray is 14,055 ft., and Shavano is 14,260 ft. All the peaks of the Sangre de Cristo are also visible, crowned with snow and rising over 14,000 ft. No one can adequately describe the beauty of the scene as these peaks burst on our view, glittering in the beautiful sunshine, so it must be left to the imagination. At the summit of the railroad we reached an elevation of 11,000 ft. and running through a snowbank entered a snow-shed where we found a thick bed of ice. The train stopped at the summit and the party descended to find themselves on the Continental Divide, where a stream from one side might reach the Atlantic Ocean, while from the other it would flow towards the Pacific ; there was a bank of snow some 25 ft. to 30 ft. deep outside, although it was about the middle of May. There was an observatory here to add to the chance of obtaining a beautiful view ; all ascended it, and some found the rare atmosphere a severe tax on their respiratory organs.

As this is not the first trip to Colorado the writers made, they will take the liberty of describing how the journey of the Mechanical Engineers could have been made even more interesting had time permitted, in other words, the traveller may do what is known as going "around the circle" by continuing to Gunnison, and after visiting various places of interest to be described, returning to Salida. If the tourist is a lover of nature, the entire journey, which is all on the narrow-gauge system, will be a constant delight. The distance by the route described will be between 700 and 800 miles, but can be extended, by taking a few side trips, to 1000 miles with much profit and pleasure, and for an additional sum of 28 dols. for the railroad ticket. The grandeur of the scenery is overwhelming, and each rugged cliff, exquisite view, or roaring waterfall preaches a veritable "sermon in stones" to the traveller.

But the query may be put here : "What is the result if the traveller is not a lover of nature ?" The answer is, that such a one may have started from home without being a lover of nature, but unless his eyesight has failed on the trip, he will by this time have become so educated that he

FIG. 38. MOUNT OURAY.

FIG. 39. BLACK CAÑON OF THE GUNNISON; THE CURRENCATI NEEDLE.

must be one whether he will or not. So then, after passing Gunnison, where the Crested Butte branch joins the main line, and noting the various mines of silver, copper, lead, and coal, together with the presence of several smelting works, the tourist passes along the beautiful river of the same name and at once plunges into the Black Cañon of the Gunnison. If this name sounds forbidding, it need not deter the traveller, for he will be well repaid for his journey. The question will at once arise as to which to admire, the skill of the engineer who laid out this line, or the nerve of the capitalist who paid for it, for the road runs for miles on a platform of rock obtained by blasting the face of the cliff, and the walls rise up frequently 2000 ft. perpendicularly, so that between these cliffs appears a little rift of sky, the bluest ever beheld. Foreign travellers have told the writers it was bluer than that so celebrated in all descriptions of the Bay of Naples. Now the train passes the spot shown in Fig. 39, where the rocks tower above to a tremendous height and terminate in a sharp peak, known by the name of the Currencati Needle. The stream at this point is very clear and beautiful, and it is said to be full of trout, but we had no time for fishing, and continued our journey twisting around curves and climbing upward until we seemed to face as many ways as some of the Hindoo idols.

A descent of nearly 1000 ft. brings us to Montrose, and from here a ride of 35 miles and a rise of 2000 ft. lands us in Ouray, a most picturesque place, though a mining town, containing some 3000 inhabitants. We may now gaze on the peaks which surround us on every side, frequently rising to a height of 3000 ft. to 4000 ft. above the town. This view, which is effectively shown in Fig. 40, is one which never wearies. Here the waters of Cañon Creek meet those of the Uncompahgre, and the peak bearing the latter most unspellable name, rises 14,233 ft. above the sea level, adding great beauty to the picture. There is, moreover, a fine cascade shown on the left of the engraving, and there are hot springs, remarkable gorges, and several caves of great beauty filled with stalactites and stalagmites. A good hotel fills the wants of the traveller, and he can readily spend time here to his great profit and pleasure.

On the other slope of Uncompahgre is Lake San Cristoval, a lovely spot easily reached from Lake City, which is on a branch running from Lake Junction between Gunnison and Montrose. The lake, with all its attractive scenery, is well shown in Fig. 41. There are two falls adjacent, Granite and Argenta, and then comes the lake, 2½ miles long, with water as clear as crystal. There are several pretty islands in the lake which

FIG. 40. THE MINING TOWN OF OURAY.

FIG. 41. THE LAKE OF CRISTOVAL.

for beauty are hard to equal. The name was given to it by an old Spanish monk in the seventeenth century. However, there is no time to linger here, for a stage ride of about two hours must be undertaken in order to reach Silverton. Much might be said of this short ride and the attendant scenery, but time presses and it can only be noted that the view from Bear Creek Bridge, where the stream falls 250 ft. in a raging torrent is most impressive and beautiful. At Silverton, a town containing about 3000 inhabitants, the elevation has reached over 9000 ft., and, as the name implies, Silverton is in the centre of a rich mining region. It stands at the foot of Sultan Mountain, which is pierced in numerous places by explorers after the precious metals, and shafts and tunnels are to be seen on all sides. There is, in many instances, a melancholy history connected with these attempts to wrest from nature her hidden stores, sometimes pathetic, and sometimes exciting, but there is no place for them in the story of a tour, and we proceed on our journey southward, for the circle we are rounding is becoming a great deal of a rectangle, and bids fair to be "squared" before the trip is ended.

The descent from Silverton to Durango is about 3000 ft., and the distance is about 45 miles, which, if the grade were uniformly distributed, would not give anything very excessive in railroading, certainly not for Colorado. The trip is through the most celebrated of the cañons of Colorado, viz., that of the Rio de las Animas Perdidas, or the river of lost souls. As this is not a particularly cheerful name to give so beautiful a spot it is called Animas for short, following the plan of the good old leader's name in the days of the Protector, which, being in full, " Praise God Barebones if Christ had not died you would have been damned Barebones," was shortened by the Royalist party, who were frequently in haste, especially at Naseby, to his last two names. On first leaving Silverton, after passing Baker's Park, the track follows the bed of the stream which roars beside it. The waterfalls are frequent and very beautiful ; then comes Elk Park, and finally, after passing the Needles, the Animas Cañon is entered. The bed of the stream is at first but a few feet below the track, which is shut in by mountains on either side, and at various points of curvature their peaks tower up many thousand feet, while the tops, crowned with snow, are sharply outlined against the sky, and seem very near, owing to the extreme clearness of the atmosphere. In Fig. 42 may be seen a good representation of this attractive spot. Passing on from here the grade becomes steeper and steeper and the track more and more curved, while two engines of

FIG. 42. Cañon of Las Ánimas Perdidas.

FIG. 43. MOUNTAIN RAILWAY IN THE ANIMAS CAÑON.

great power are now necessary to haul the train. The road in spots seems literally fastened to the side of the cliff, and one may gaze from the car window nearly 1000 ft. below where the river looks almost like a white cloud, so flecked is its surface with foam, for its descent is rapid and the rocks and cascades quite numerous. Here and there, clinging to the sides of the cliff in crevices, which look as if they could not furnish enough earth to support a daisy, are to be seen hemlocks and pines. The picture shown in Fig. 43, gives a fair idea of the engineering feats accomplished, and a faint notion of the beauty of Animas Cañon. After whirling along for some miles the train descends into a beautiful valley, the traveller passing on his way, Trimble Hot Springs, whose mineral waters are alleged to have wonderful curative powers, and which the writers did not taste, as they were in excellent health. The road now follows the river, and alongside are fertile fields under cultivation, the many farmhouses of excellent appearance affording evidence of prosperity. After passing nine miles of attractive scenery we are landed at Durango.

A ride of about 170 miles from Durango brought us to Antonito, and in this journey we passed through the famous Toltec Gorge and Toltec Tunnel. Here one may look upward to cliffs 1000 ft. above, and gaze downward some 1500 ft. Although the tunnel is but 600 ft. long, it was, as may be supposed, a most expensive work. The rock is extremely hard, and, in fact, to-day the roof of the tunnel needs no support.

The road is sustained here by a solid iron trestle set into the rock ; this is to be seen in Fig. 44, and lest the reader overlook the train in his admiration of the picture, which is taken from the bottom of the gorge, his attention is directed to a small object in the upper right-hand corner, which represents the train about to enter the tunnel, the latter being a small black spot, at first glance seeming to be a spot of ink, while the train might be passed over as a smear of the printer. At this place there is a beautiful memorial monument to the late President Garfield. We must now rapidly transport our traveller to Salida, much as the scenery tempts one to enter upon detailed and illustrated descriptions, and take up the story of the engineers where we left them at the summit of the Marshall Pass. The ascent had required two engines of the four-coupled consolidated type with 20-in. by 22-in. cylinders and 35-in. drivers. We returned the 25 miles with but one engine, and that evening found the party at the Grand Concert Hall in Salida, where the following unique signs attracted our attention : " No loud whistling or yelling allowed," and another, " Any person caught

FIG. 44. THE TOLTEC GORGE.

throwing anything from the gallery to the floor below will be at once
expelled from the building and will not be permitted to enter it again."
After a pleasant evening the party retired to their train, and awoke to find
themselves *en route* to the Tennessee Pass, having passed Leadville at an
elevation of 10,200 ft. during their slumbers without knowing it.

One of the great charms of this method of travelling was shown here;
we had not time to observe all the wonderful scenery in detail, so the finest
was selected, and by remaining at night on the side track instead of
pursuing our journey, we were enabled to see all the best by daylight.
At Tennessee Pass we reached 11,100 ft. in elevation and had, later, a
view of the celebrated Mountain of the Holy Cross, so called because near
the summit a cross is marked out by the rock formation on a grand scale.

In this pass we were shown the Butler Mine, named from the General,
and the timid ones examined their souvenir spoons, and rejoiced to find them
intact. We soon came to other mines, in some places situated so high on
the mountain side that the ore had to be lowered by wire ropes, and it
would seem to the uninitiated as though the miners must go to their work
by means of balloons ; soon we entered the Cañon of the Grand River, and
saw on one side of the stream a large mass of lava, evidently thrown up by
a volcano extinct for hundreds of years. The rocks became varied in colour
and reminded the spectator of architectural works on a most enormous scale ;
castles with towers hundreds of feet high, fortresses whose outer walls would
defy Titans, and every fantastic shape which can be imagined. The cañon
is narrowed at its entrance, which is called the portal, and a view of it
appears in Fig. 45, which gives the best idea of it possible without
reproducing its many colours. Although the morning was bright, the
valley seemed more and more shut in, and the scenery assumed a sombre,
gloomy look, the only bright feature being the sparkling water flowing
alongside the track. The foliage had likewise a funereal appearance, and
the railroad even felt constrained to hide itself on three separate occasions,
the first of which is seen in Fig. 46 ; but suddenly the hills fall away on
either side, the river expands and dances in the sunlight, the foliage takes
on its wonted green, and all nature seems to rejoice with us as the train
halts and the party alights at Glenwood Springs, where we find our
elevation to be but 5768 ft.

The whole party proceeded to cross the river to the springs, feeling
sure that " cleanliness is next to godliness," and desiring a suitable
preparation to fit them to enter the city of the Latter Day Saints, which we

FIG. 45. THE PORTAL; GRAND RIVER CAÑON.

K

FIG. 46. TUNNEL ON THE GRAND RIVER CAÑON.

expected to do on the following day. In the shortest time all were busily sporting in the open air natatorium, shown in Fig. 47, where the water, surcharged with soda, bubbles up at a temperature of 120 deg. and at the rate of 2000 gallons per minute. The novelty of bathing in a warm spring and being able to look up at the snow-capped mountains on every side, will not soon be forgotten, nor will the chilling effects of the fountain of spring water shown in the picture. The change from the warm water to the cold was varied in its effect on the bathers. The men gave a stolid sort of a grunt and swam away, but whenever there was a shrill feminine shriek we knew one of the ladies had swum into this locality.

FIG. 47. GLENWOOD SPRINGS.

In the afternoon the party was on its westward way, and soon we reached a country so utterly different that it seemed almost incredible that only so short a distance separated it from the rocks of the Royal Gorge. The flat plains, beautiful with wild flowers of every hue, are surrounded by great soda buttes of strange and fantastic formation. Now, by a stream, these are comparatively low and riddled with the holes of wild birds, and again they rise on high in the exact shape of a mighty palace of the ancient Aztecs, or in the various towers of a great castle. As we passed one little village we saw a train of galloping horses, and behind them cantered a real cowboy with broad sombrero, and a lariat in his hand.

The plain before him, yellow with the blooming sage, its only plant, was
bounded by a line of lofty buttes like the wall of a city; where the sun
struck them they were pale yellow, while the shadows on the distant ones
were deep purple, and behind, the low clouds drove across the blue sky in
heavy masses of white and grey.   At 6 o'clock we came again on the
Grand River.   Here the buttes plunge sheer to the water's edge, their
deep red casting a bright reflection in the clear broad stream ; strange and
wonderful in shape, they resemble now a Greek temple, and now the spires
of a cathedral, or a solemn row of colossal statues, each one separated
distinctly from the next, and taking the mind to the days of Egypt and
the Temple of Thebes.   Gazing back through the curving and opening
vistas, the red glow of the massive cliffs contrasts with the vivid light
green foliage bordering the river.   It has been suggested that this cañon
was once the bed of an enormous river which rolled its deeps between
these lofty cliffs, and that their seams and furrows were worn by the rush
and turmoil of its waves.

After passing Azure Cliffs, so named from the colour of the clay
formation, we entered Castle Cañon.   We had now crossed both the Green
and the Grand Rivers, but these unite some miles from here to form the
Colorado.   This cañon is full of weird and wonderful scenery ; in one spot,
on the summit of a towering cliff, was a mighty dome, large enough to
cover the Columbian Exposition grounds at Chicago, and history can
furnish no greater simile ; in another place a rock rises thousands of feet,
having little spurs like minarets on all sides, and taking the form of a
gigantic steeple to some sunken church.   At the entrance to this cañon
are two large sandstone rocks 500 ft. high, standing like sentinels, leaving
scarcely room for the tiny railroad track to enter.   The grade rises rapidly
again, the track passes through Castle Gate (Fig. 48) and enters the Price
River Cañon.   The two pillars, about 500 ft. high, are of a deep red, and
this is set off in contrast by the pines which grow at the edges of the base.
The views in the cañon are still on the magnificent order, and every eye
was strained to see them till night, but even then it was not lost to us,
since a full moon kept up the interest, throwing a sort of spectral effect
upon the scene, and late that night, on raising the state-room curtain, we
could see the level stretches of sage brush, looking as though covered with
a heavy frost, the great buttes standing out black against the silver clouds.
From the gateway to the Cañon of the Wahsatch Mountains the grade
rises to 227 ft. to the mile, requiring two engines for one train, and the air

FIG. 48. THE CASTLE GATE; PRICE RIVER CAÑON.

was so pure and exhilarating that the spirits of the party rose like cream
on new milk, one of them being heard to declare in his ecstasy, "I declare
I feel good enough to join the Mormon Church," and the opportunity was
soon to be afforded him, for the next morning found us at Salt Lake City,
and in the presence of that strange body of people. The approaches to
this place from the cañon just described were novel and attractive. In the
early morning we passed several "dug-outs," which were nothing but
holes in a side hill, with timbers placed to hold the earth up. At one of
these openings we saw at least seven people, and near by was evidently a
boarding-house, also a "dug-out," with the entire family at breakfast;
the host being somewhat cramped for room had considerately put the box
serving for a chair outside the building, but his head and shoulders were
inside, and he bent fondly over the table constructed of several boxes.
The sign over the door was startling and probably attractive. It read:
"Meals 35 cents; no charge if not satisfactory." Unfortunately the
method of determining satisfaction was not stated. Perhaps it was similar
to that practised by a Western barkeeper when a rough-looking lot of
miners were drinking with a tenderfoot. As they leaned against the
plank resting on two barrels which formed the bar, the barkeeper took
their orders, it was uniformly "whisky" till the tenderfoot mildly asked
for "sherry." The barkeeper's revolver was out in a minute. "I won't
have no tenderfoot putting on airs with my regular customers," said he.
"You meant whisky, I reckon?" "Certainly, certainly, sir, of course I
meant whisky; never drank sherry in my life." Alongside the track for
many miles was a large fresh-water lake called Utah Lake, which
added greatly to the beauty of the scenery. Herds of cattle were to be
seen, and later we passed a typical "prairie schooner." This was a wagon
with a white top cover of sailcloth, and a door formed in the side with two
small glass lights in it, while through the top of the cloth projected a
stove-pipe. This was really a luxurious type, and as the door was open,
we saw a woman cooking breakfast; evidently her presence was the
explanation of such unwonted style. Brick kilns were to be seen at short
intervals, and even the smallest houses were of this material. A wooden
house was a luxury, and we only saw a few of them, but each was in
the midst of an attractive-looking lawn, and bearing evidence of prosperity.
There were many fruit orchards in view, all in full bloom, and the varied
colouring of the blossoms, and their fragrance borne on the air, refreshed
every one. The country being the promised land to the Mormons. was

indeed fertile, and scriptural names such as Moab, Jerusalem, and the like were frequent; presently we passed over the Jordan, but experienced none of the ecstasy described by enthusiasts. Possibly, if the truth was known, the reality was not all fancy depicted, and they may have felt like the old negro, whose nightly prayer was " O Lord, send the angel to take this nigger to heaven." But when some jester knocked on the door and announced himself as the angel come for the purpose of fulfilling the request, the darkey in great fright roared out " Go long, you old fool, that nigger been dead this three weeks."

The Jordan is crossed by a Mormon bridge which was erected without the use of a nail, that desirable object not being among the Mormon stores at the time of its erection. We also passed charcoal huts, and later some-

FIG. 49. BRIGHAM YOUNG.

thing which may interest foreigners, a large beetroot sugar refinery, that industry having taken quite a hold in this section, and it is understood, with considerable success.

On reaching **Salt Lake City** we expected to meet long-haired, wild-eyed men of stern, forbidding countenances; on the contrary, a short pleasant-faced gentleman met the party and we started under his guidance for the Mormon Temple.

The speculation at once became rife, especially among the ladies, as to whether he was a Mormon, and if so, how many wives he had and the number of his family, but no one put the question direct. When the party alighted from the electric car, taken from the Railroad Station, they were the objects of considerable attention, and in return they looked interestedly at the people. Naturally they assumed all were Mormons, and yet it is doubtful if any were, for the Gentiles are very numerous.

Of course Mormonism centres around Brigham Young, " who, though dead, yet speaketh," and what was pronounced by one of his friends to be an excellent likeness is given in Fig. 49.

The first object of interest visited was the new temple shown in Fig. 50. This was commenced in 1853, and thus far has cost 6,000,000

FIG. 50. THE MORMON TEMPLE.

dols. It is rapidly approaching completion, and if the Gentiles will only visit it enough, its progress is sure, for an admission fee of 25 cents is exacted. In our case it was gracefully waived, and then we were quite sure our guide was a Mormon, and so he was ; he was also an English gentleman who had been here some twenty-five years, very kind and desirous of showing us all hospitality. It was rumoured he had seven wives, but this may be an error ; no one asked him, but he volunteered the statement he had twenty-four children, one of them six months old, and regretted we could not visit his wife (he certainly used the singular form) and see the infant. If all the fold were as pleasant as he was, they must be charming to meet. The temple is of white marble, or, to be exact, of white granite which is quarried fifty miles distant at Cottonwood Cañon. The dimensions are

200 ft. by 100 ft. with a height of 200 ft. The four towers are each 220 ft. The one on the right-hand corner is surmounted by a colossal gilt figure of an angel with a trumpet at its mouth, and very large electric lights are distributed about it. The effect at night must be grand. The walls of the building are 10 ft. thick and the interior will be handsomely decorated. The situation is a commanding one and the structure can be seen fifty miles distant.

A short distance from the temple is the tabernacle, two views of which are given ; the exterior is seen in Fig. 50, and the interior in Fig. 51. This enormous building, made of iron, glass, and stone, is 250 ft. long

FIG. 51. INTERIOR OF MORMON TABERNACLE.

150 ft. wide, and 100 ft. from floor to centre of dome. The design was revealed to Brigham Young by an angel, who was not only a good judge of architecture, but also a close student of acoustics, for by standing about in the centre of the tabernacle, a whisper made at the railing to the right of the organ, and in front of the open door on the left, can be distinctly heard ; and, more marvellous still, a pin dropped there some 3 ft. or 4 ft. above the railing was heard distinctly by our entire party when seated on the benches near the centre. There are twenty doors 9 ft. wide, opening outward, so that the building can be filled or emptied with great facility. The organ at the background is a beautifully made and exquisitely sweet-toned instrument. It was constructed by Mormon workmen, from the native woods, at a cost of 100,000 dols. In height it is 58 ft., it has 57 stops, and

L

2648 pipes.   The choir gather on either side, and number 500.   The president and his two counsellors occupy a conspicuous sofa, while the other dignitaries occupy places on steps according to their rank, the subordinate ones using inclosures to the right and left.   The hall will seat 13,462, and the seats are free ; nor are Gentiles excluded.   The service is largely of song, together with reading from the Mormon Bible, and then there may be a short discourse from any of the officials who desire to speak.   They have a Communion service, using the water of the Jordan in place of wine.

A little beyond the Tabernacle is the Assembly Hall, shown in Fig. 52. This is also of granite ; it cost 90,000 dols., and will seat 2500.   On the

FIG. 52.   THE ASSEMBLY HALL, SALT LAKE CITY.

ceiling, which is divided into sixteen panels, is a grand fresco illustrating the history of the Mormon Church.   One scene shows the angel delivering the golden plates to the founder of Mormonism, Joseph Smith.   These were written in a cipher that only Joseph could translate, and contained the foundation of what is now the Mormon Bible.   The reader will understand the Mormons accept the Gentile Bible as far as it goes, but claim a new and later revelation.   Acoustic properties are again a feature of the construction of this building, and many eastern architects might study the example with profit to themselves and pleasure to their audiences. East of this is the Mormon Tithing House and printing office, Brigham Young's residence, the Lion House, and the Beehive House, so named from having a carved lion on the one and a beehive on the other.

Here is also a pretty school-house, where his seventy-eight children attended. A man with such a family was certainly entitled to a private school. The late Artemus Ward use to tell of a flourishing young ladies' seminary with thirty-five pupils which came to grief by a sly Mormon elder eloping with the whole of them. Eagle Gate, which spans the street, has four piers, with a carved eagle over the centre resting on supports running to each pier.

As to the Tithing House, a word may be said. When the Mormons were in full possession, the payment of one-tenth to the Church was rigorously exacted, but now, of course, it is voluntary, although my informant told me all true Mormons paid it, and if he made 200 dols. he at once sent 20 dols. to his church. Was it any wonder that with such a revenue, and in the midst of such a fertile region, the Mormon Church grew enormously rich ? Brigham Young also built a theatre in 1862, which is to-day an imposing structure. It is 175 ft. long by 80 ft. in width, and will seat 1500. It is built of adobe and finished with granite. The style of architecture is Doric.

A great source of revenue to the Mormon Church is the Zion Co-operative Store. Our party was taken through this enormous establishment, which would be a credit to any city in the world. The first floor, stored with dry goods, will compare favourably with the best New York and London stores in attractiveness and tasteful arrangements. The writer on seeing the other floors was reminded of the Army and Navy Stores in Victoria-street. The large shoe factory was another feature, but whether the Mormons make the "shoes of righteousness " or not, they certainly turn out a good article. Everything, apparently, can be purchased at the store, from a triple-expansion engine to a thimble, and from a cotton pocket handkerchief to a Paris dress. The aggregate business exceeds 6,000,000 dols., and some of the party who had dealt with this establishment said all accounts were paid promptly on the first of the month.

The territory of Utah was almost unknown until 1833, and was considered as a desert. That year Captain Bonnĺville explored it to some small extent, and in 1842 to 1846 Colonel John C. Fremont made further explorations ; there were then a few tribes of Indians as its inhabitants. In 1847 Brigham Young and 142 pioneers came into the region. Their first action was to begin the irrigation and planting of this arid waste. Young certainly was a remarkable man and gifted with great foresight. He also knew how to turn any unusual event to his own benefit. For instance, when there was a sudden advent of sea-gulls, which devoured an army of black crickets that threatened to devastate the crops, this was set down as a miraculous interposition of Providence.   The gold hunters flocking to California in 1849 started up a " boom " in Salt Lake City, and in 1850 the territory was organised under an Act of Congress, and Brigham Young was appointed governor.   There were then 30,000 people in the territory and 5000 of them at Salt Lake City.   The territory continued to grow, and in 1870 the last rail was laid and the last spike driven, of the Utah Central Railway, by Brigham Young.   With this railroad came the usual number of new settlers, and Mormonism was doomed in 1880 ; the population of Salt Lake City was then 20,678, and in 1890, 46,259, while to-day it is upwards of 55,000.   The valuation of property has risen from 16,600,000 dols. in 1889 to 54,350,000 dols. in 1890.   It has also 65 miles of electric railways.

The climate of Utah is extremely mild, and the air is fresh and pure. The boast of the territory is the great age of its inhabitants ; one of the common mottoes is, " We believe it is a duty to live past seventy," and under this are frequently placed a row of photographs of well-known inhabitants, some of whom are ninety, and all bearing the proud look due to a consciousness of duty well performed.   The following extract serves to confirm this statement : " ' Old Folks' Day ' is a Utah Mormon institution, which might well be made national in its scope and observance.   It was established by Bishop Hunter, of the Mormon Church, who died at the age of ninety years.   It comes on June 22, and is observed as a general holiday.   An excursion is given to people of seventy years and upwards, winding up with a banquet, a dance, and a general distribution of presents. In 1887, when Salt Lake City had but about 30,000 population, she sent 750 of these ancient jollifiers over the Rio Grande Western Railway to Ogden.   Of the number, 112 ranged in age from eighty to ninety-seven. A seventy-year old papa, trundling a baby chariot, with a springy tread of a young game rooster, is no uncommon sight on city street or country road." The country seems prolific in every way.   For instance, one man raised in 1890, on 20 acres of land, 1920 bushels of oats, averaging 96 bushels to the acre, and the year previous he had raised 104.   Another man raised in 1890, 90 bushels of barley to the acre ; 112 bushels of corn to the acre were also raised, and, again, 947 bushels of potatoes, while at another point a farmer cleared 1200 dols. per acre on strawberries ; 7 tons of clover per acre is not unusual, and they frequently cut four crops of it in a season.   The following extract from the last Chamber of Commerce report may be cited at this point :

" The earth is absolutely wanton in fecundity. Rye yields an average of from 60 to 70 bushels to the acre ; turnips from 400 to 600 bushels ; carrots, from 700 to 1000 bushels ; apricots, 350 to 500 bushels; peaches, from 500 to 700 bushels ; apples, 450 to 600 bushels ; pears, 500 bushels ; plums from 300 to 400 bushels; blackberries, raspberries, currants, and gooseberries, from 300 to 350 bushels to the acre, and everything else in like profusion. Cherries grow wild in great abundance. Hops are indigenous to the soil. Nectarines flourish everywhere, and figs are raised in the southern valleys. Cotton grows luxuriantly in the lower counties, and a cotton mill established by the Mormons at Washington has long been in successful operation. It uses about 75,000 lb. of cotton yearly and manufactures good domestics." Nor does the human race claim any exemption to this law. " Be fruitful and multiply " is the rule everywhere, and as an evidence of this it may be said that at Pleasant Valley, where the population numbers 2300, there are over 800 children of an age to attend school and 378 of younger years.

As a mark of courtesy, the officers of the society were presented to the rulers of the Mormon Church, President Woodruff and the two counsellors, President Cannon and President Smith, the latter being a nephew of the celebrated Joseph Smith, founder of the order, who was shot by a mob in Nauvoo, Illinois, when the people of Illinois rose and expelled all the Mormons from their borders. President Cannon is about ninety years old, but looks like a man of seventy. All of these gentlemen were well-preserved, stout men, in a most excellent physical condition, with fine intelligent faces and a look of shrewdness which impressed the spectator with their ability to deal with almost any problem in a most successful manner. Knowing the party to be eastern men and thoroughly loyal to the Government, they spoke in the kindest way of the administration in a rather soft deprecatory manner, full of sadness that they should be so misunderstood and maligned, but to the writer's mind there seemed to be an undercurrent of intense hatred, bitter and uncompromising. Nor is it to be wondered at. Their claim in general is, that they found an absolute desert, and bent their skill to its improvement ; that they solved the irrigation problem, collected people of their way of thinking, and by economy and prudence have created a populous and fertile land which they think belongs to them ; that when their efforts had made this desirable, the cupidity of the Gentiles had seized upon it, passed laws against their church holding the property it has created, and, as they believe, are fairly entitled to; attacked their religious beliefs by prohibiting polygamy, and reduced them to a condition of vassalism by sheer power of arms.   There is something to be said on their side beyond a doubt, and that they are a frugal and industrious people no one can deny; that they have solved the Indian problem far better than the nation at large is also true, perhaps because they had a lower order of red men to treat, but certainly to a great extent because they were kinder in their methods.

The only answer to all this is the general statement that no nation, least of all a republic, can tolerate within its borders a separate system of government which owes no allegiance to the general government; that polygamy is abhorrent to all nations of this age, and is opposed to good citizenship; and that Mormonism, wherever it was against the public good of the nation, was by that very act a doomed institution, and could not be tolerated; that from a purely religious standpoint, independent of any conflict with the laws of the land, it is not disturbed so far as it is a belief; and that if they used their power and wealth to oppose the Government, as they undoubtedly did, they were in a state of rebellion, and by that very act forfeited any claim to consideration.

This discussion on Mormonism may be most fittingly closed with the grave of its greatest apostle, and probably its most shrewd and competent leader, which is shown in Fig. 53, inclosed by an iron fence.   The cemetery is located pleasantly on a side hill, and but a short distance from the temple whose corner stone he laid, but which, like the religion he founded, he was never to see complete and established.

In the afternoon the party visited the Great Salt Lake.   This wonderful body of water, whose only parallel is the Dead Sea, to which the Mormons love to liken it, contains 18 per cent. of solid matter, mostly salt and soda.   It was once as large as Lake Huron, and is now 100 miles long, with an average width of from 25 to 30 miles.   Hundreds of thousands of tons of salt are made by natural evaporation along the shores of the lake, and at one place near Salt Lake City, a windy night never fails to pile up many tons of soda, eliminated by the movement of the waves.   Four large rivers pour fresh water into it without raising its surface or diminishing its saltness.   It was formerly supposed that no creature could live in its waters, but lately scientists have discovered there a few of the lowest orders of microscopic organisms.

It was a great disappointment to us on our arrival at Garfield Beach, to learn that the weather was far too cold to admit of bathing, for the first bath in the Great Salt Lake is said to be an exciting event. The human body cannot sink ; you can walk out in it where it is 50 ft. deep, and your body will stick out of it from the shoulders upwards ; you can sit on it ; men lie and smoke with their arms crossed under their heads. But the great difficulty is to move gracefully, to keep from turning somersaults and remaining heels up. The water is said to contain powerful medicinal virtues.

The likeness of Utah to Canaan, which led Brigham Young and his 20,000 Mormons to settle there, is indeed very striking when the two maps

FIG. 53. THE GRAVE OF BRIGHAM YOUNG.

are placed side by side. But with the similarity of contour the likeness ends unless one considers the methods of the old Israelites, and there indeed a similarity may be discerned. From a purely artistic standpoint also the Salt Lake is interesting ; the waters are of the most lovely and varied shades of blue and green, and the surrounding country is fertile and beautiful. We left the lake as the sun was setting, and looked back at it as we steamed away from the city, gazing across plains white with salt to its placid waters, over which was a pale green sheen ; the high purple cliffs stood like a gateway in the west, and on the horizon a broad band of gold cast a stream of gold upon the waves.

The party soon reached Ogden, and were delayed there from bad management on the part of a railway official, who sent out the regular train, which made frequent stops, and held back our special, which only stopped for fuel and water, until the other train was well out of our way. However, we did start finally, and were soon going toward the home of the setting sun at a fair speed, having a proud consciousness from our pleasant impressions of Ogden that it meant something to be an American citizen. Of course up to this point may be noted by examining the map shown in Fig. 54.

By morning we were in the State of Nevada, and while we sat at breakfast we saw a solitary Indian hunter, gun in hand, striding over the prairie, evidently on a hunt, possibly seeking his breakfast. Chinamen were also to be seen, and their names were as curious as their personal appearance. Wun Lung—Wun Lung of Carlin, Nevada—was rather the most startling ; and yet anyone who examined his smiling countenance would not think he experienced any sense of deprivation. This may be due to the extremely pure air, so that one lung here is as good as two anywhere else. Our Kodak was fortunate at Carlin in obtaining a picture of an Indian woman. The Indians are extremely difficult to photograph, especially in the case of the women, since they have a superstition that whoever possesses their likeness can control the original, and may at any time cast a spell on them. Instantaneous pictures are the only kind attainable, and these can be procured by diplomacy, although one has to be very quick, and very sly even then. They would cover their faces with their blankets as soon as the Kodak appeared, and in some instances would cry from fear. The more civilised would at times permit pictures to be taken at two bits each (25 cents), but these cases were not numerous. Journeying onward, we next had our attention called to the Humboldt River, which in this section literally disappears, although it is a good-sized stream, and passes away, its waters being absorbed into the earth.

During the afternoon we passed the alkali plains, which usually cause so much discomfort to travellers from the poisonous dust that arises and permeates the air and the train, entering at every crack ; but here again we were most fortunate, for the falling rain had drenched the earth, and there was no discomfort whatever from the dust. At about 2 P.M. we stopped at a little town, and it would be hard to imagine a picture more utterly dreary and forsaken-looking. A few dirty white houses on very broad clay streets which had neither grass nor trees, while the level prairie stretched away in all directions, utterly barren of any object. On the platform half a dozen squaws with gorgeous yellow and blue shawls about their shoulders and around their copper-coloured faces, attracted the vigilant

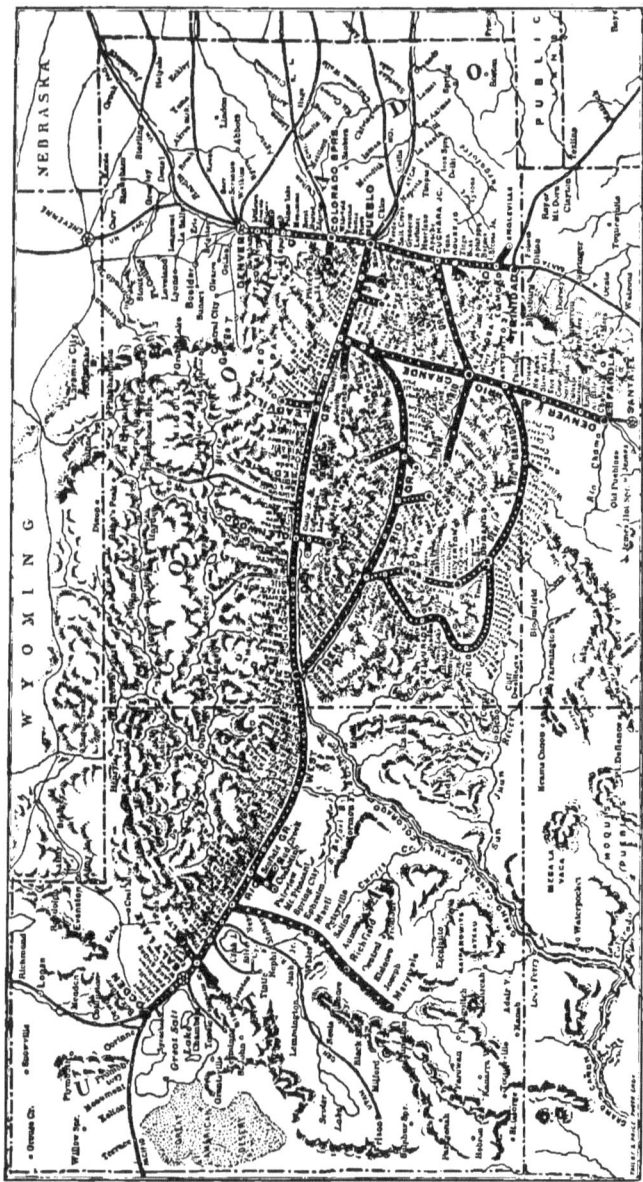

FIG. 54. RAILWAY MAP BETWEEN DENVER AND OGDEN.

Kodak fiends, who left the train in large numbers hoping for many snap shots. But the gentle savages were horribly frightened, and, gathering in a tight wad of shawls, protested in noises like cat-calls. One woman had a papoose strapped on a board, and the ladies paid as much as 25 cents. for a peep at its tiny face, though she refused to be photographed, although no less than 1 dol. was offered for the privilege.

The cowboys, who stood around in groups and grinned at us, were picturesque in their broad felt hats and big boots. One great creature with an immense yellow beard attracted the notice of an inquisitive lady, so she said :

" How long have you been in this place ? "

" Twenty-four years " was the reply.

" Why, whatever did you find to do here that long ? "

" Oh ! I drink whisky mostly," said the giant, adding frankly,

" I don't drink it like you drink, I just swalleys it right down."

On passing Reno we soon entered the State of California, and presently began to go up the slope of the Sierras, running that night through more than forty miles of snow sheds. At the summit the elevation was 7017 ft., and the grade had been 210 ft. to the mile. About midnight, after passing through a long snow shed, the train was stopped alongside a snow bank 7 ft. high, and placed on a side track till the morning, that the party might enjoy the descent on the Pacific side by daylight.

It was at this point we were met by our future hosts, who had come in a private train nearly 200 miles to welcome us, and with them they bore the fruits of the country. They were evidently very wide awake, for we were brought out of our beds at 5.30 A.M. and presented with fruit and flowers, and an informal reception was at once organised in our combination car. Each of our ladies was presented with a beautiful souvenir spoon made from the silver of Nevada; it was shaped like a miner's shovel, and on the cross-handle was a bear in relief, that being the emblem of California. Flowers were showered upon us, and every part of the train was decorated ; even the engine did not escape. The occasion was one of great gaiety, and the warmth of our welcome made a most delightful contrast to the cold of the snow and ice which surrounded us. Attaching their train to ours the descent of the Sierras began, and the panorama from the windows seemed to be like the changing of the seasons of the year, as we passed gradually from the ice and snow of February to the early spring. The hill-sides commenced to throw aside their wintry garments and to put on the cheerful

green of April. The air grew milder and the valleys below began to stretch
away in luxuriant vegetation. Here and there a flash of gold along the grass
showed the presence of yellow poppies, mingled with masses of bluebells and
varied shades of clover and forget-me-nots. Next, as we whirled downwards,
we saw peach orchards in full bloom, and the various smaller fruit trees,
with many-hued blossoms, forming a most delightful contrast of colour, until
at last appeared the fig and orange trees. We were, in truth, in the Golden
State, and taking advantage of a halt, the entire party ran upon the station
platform like children, romping and shouting and inhaling with delight, deep
breaths of the fresh perfume-laden air, while near by was a jolly crowd of
labourers singing and gathering cherries. We contented ourselves, however,
with great bouquets of wild flowers of singular beauty. Later in the
morning we reached the town of Colfax. One great industry at this
place is the transport of freight by wagons from the railway to the interior.
Sacramento, the capital of California, was reached about noon. But before
arriving we received a telegram asking us to take part in the laying of a
corner stone by the Knight Templars, which had to be declined. The first
visit at Sacramento was that to the Southern Pacific shops, the party
being escorted by the superintendent of motive power. A beautiful souvenir
programme printed in blue and gold was given to us, and the following
quotation from it will best describe these shops and the city :

" Sacramento, which figured so prominently in the early days of
California, is the capital of the State, with the glorious clime, the sunny
slopes and fertile valleys. It is situated in about the centre of the richest
and most productive valley of the State, the valley of the Sacramento ; and
in full view of the picturesque and lofty Sierra Nevada mountain range.

" Following the days of the discovery of gold in this State, it became
the chief distributing point of supplies and mining machinery to adjacent
mines, and is now the principal shipping centre of the fruit-growing
industries. Three-fourths of all the fruit shipped from this State each year
is sent from this city. It has a population of some 29,000, and is steadily
increasing. The residence portion of the city is beautifully laid out and
abounds in many handsome dwellings and buildings, the most important of
which is the State Capitol, situated in the heart of the city on a public park
embracing ten blocks. It is one of the finest structures of its kind in the
Union, and was erected at a cost of 7 million dollars. The State Printing
Office, the State Agricultural Exposition building, and an Art Gallery are
among its chief attractions ; the latter contains the finest collection of Art

paintings on the Pacific Coast, and was donated to the city by that most estimable and philanthropic of ladies, Mrs. E. B. Crocker. The city is somewhat extensively engaged in manufacturing, is lighted by electricity, and has excellent street and railroad facilities.

"Here are also located the main shops of the Southern Pacific Company which cover an area of 42 acres, of which 17 acres are occupied by buildings employing from 1800 to 2000 men continually, with a monthly pay roll of 125,000 dols. These shops are principally devoted to the maintenance and rebuilding of rolling stock of the 4500 miles of railroad constituting the Pacific system of the company, consisting of 731 locomotives, 931 passenger train cars, and 15,712 freight and miscellaneous cars; also the general repair work of the floating equipment of the company, consisting of 26 ferry and river steamers, tugs, and barges. The works are conveniently arranged for the rapid and systematic production of the different classes of work. The machine shop is well equipped and has a travelling crane capable of lifting the heaviest locomotive. The forge shop has a 30-ton jib crane, numerous steam hammers, and can handle forgings up to 20 tons in weight. The output of axles ranges from 40 to 50 per day. Here are also produced the steel brake beams used on the road, which are forged from old steel rails. The rolling mills have an annual production of 12,000 tons. The wheel foundry has a capacity of 300 wheels per day, and the average daily melt of the cupolas is about 60 tons. In the car shops, constructions, repairs, painting, and upholstering for all classes of railway cars are well provided for. The forests of California, Oregon, and Washington, furnish the supply of lumber used here. The paint shop has a capacity for handling 21 passenger cars and 8 large Pullmans. In addition to the general railway repair work, the output of these shops for 1891 was as follows :

"Chilled car wheels, 29,854 ; rolled iron, 12,753,997 lb. ; iron castings, 8,154,878 lb. ; brass castings, 265,295 lb. ; journal bearings, 241,764 lb ; phosphor-bronze castings, 40,746 lb. ; Babbitt, 284,374 lb. ; track spikes, 2,460,900 lb. ; track bolts, 627,596 lb. ; nuts, 678,377 lb. ; angle plates for track, 3,981,668 lb. Of new work there were built in late years some 63 locomotives, ranging from 8-wheeled passenger to 14-wheeled or Decapod locomotives.

"The building of cars for last year was 547 new freight cars of 30-ton capacity to replace worn-out small capacity cars. Everything used in the construction of this new work was manufactured at the Sacramento shops. The shops also have a well-organised fire brigade composed of 40 employés, with an equipment of 5 fire pumps and 5 hose carts ; and regular drills are had every two weeks."

We went afterwards to the Crocker Art Gallery, and were quite surprised at the number of the paintings and the care and skill displayed in their selection. Many of the choice works, not only of the old masters but of the great modern painters, were to be seen, and our only regret was the brevity of our visit. The day was quite warm and the party was taken

FIG. 55. THE CAPITOL, SACRAMENTO.

to a large brewery, not only to see how beer was made, but to find out how good it could taste, thus gratifying two senses. After this we drove around the city viewing the many beautiful residences and passing through the park ; a visit to the Capitol (shown in Fig. 55) completed a delightful day. The interior of this building is as beautiful as the exterior. It is rich in decoration, but nowhere is it overloaded with ornamentation. Many fine portraits of the governors and statesmen adorn the walls, some of them early pioneers in this great State. The grounds are also attractive and tastefully laid out.

That night, while slumbering peacefully, we were transported as though by the wave of a magician's wand from the ordinary earth to fairyland, for we waked to find ourselves at Monterey, and in front of the celebrated Hotel del Monte. The writers feel their power utterly inadequate to picture the beauty of that morning wakening, and can only hope the illustrations, which lack the various colours, will assist the reader's

FIG. 56.   HOTEL DEL MONTE.

FIG. 57.   THE ARIZONA GARDEN.

imagination. Descending from the train we found ourselves surrounded by a forest of pine and oak trees, and following the pathway for some distance suddenly the scene (see Figs. 56 and 57) burst upon our eyes. Fig. 56 shows a part of one side and the greater part of the front of the hotel, and was chosen so as to give the best idea of this beautiful place, which is situated in a park of 120 acres, and surrounded by trees whose age runs far back into the last century; nor is the general location of the town less picturesque, for it is situated on a promontory having the Bay of Monterey on one side, the Pacific Ocean in front, and Carmelo Bay on the third side. Practically, for these bays are but arms of the ocean, it may be said to be surrounded on three sides by the Pacific. It is difficult to single out any one spot in the grounds of the hotel as more beautiful than any other where all are so beautiful, yet the writers feel that special attention may be called to the Arizona Garden (shown in Fig. 57). If but a portion of the many colours there could be produced by the photographer's art, the reader would feel that enthusiasm has not in any sense carried too far appreciation of the beauty of this place. As we followed the winding paths into the shade, one lovely vista after another opened before our eyes: now a cluster of beds brilliant with the great blossoms of white, pink, red, and even orange roses; now a tropical garden or broad sweep of green terrace to the Laguma del Rey, in whose centre rises the wavering jet of a fountain. The soft air is laden with the perfume of flowers; on one side of the house is a portico draped with huge white roses, and the wall beside it is covered with mammoth heliotrope.

Sea bathing is enjoyed all the year round, and not far from the hotel is a handsome pavilion, designed for those who find the outdoor bathing too bracing. This contains four immense swimming tanks of varying temperature, and is roofed with glass, floored with marble, and decorated with numerous palms. When there is added to this a beautifully appointed hotel, with every provision of comfort, good service, a great variety of tempting food, well cooked, and appetisingly served, is it a wonder our party found it most difficult to leave the spot? Or that it remains a distinct feature in the many pleasant recollections of this attractive journey?

In the afternoon, many of the party visited the quaint old town of Monterey, distant about a mile from the hotel. The mind at once ran back to that great and noble struggle carried on in these crooked streets and lanes, nearly three centuries ago, and we may quote at this place from the historical record.

"From the earliest period of California's history," says Harrison in his "History of Monterey County," "Monterey has been conspicuous as the objective point of navigators and explorers, and the arena where were enacted many of the important political and historical events of the county. As early as 1602, Don Sebastian Vizcaino, sailing under instructions from

Philip III. of Spain, entered Monterey Bay, and landing with two priests and a body of soldiers, took possession of the country for the king. A cross was erected and an altar improvised under an oak tree, at which was celebrated the first mass ever heard in the land now known as California. The place was named in honour of the Viceroy of Mexico, Gaspar de Zuniga, Count of Monterey, the projector and patron of the expedition. The departure of this expedition returned the place to its primitive condition, and the silence in its history was not broken for a period of 168 years. When Father Junipero Sera, president of the band of Franciscan missionaries sent to the coast in 1768, was planning his work in California, the most cherished object of his expedition was the founding of a mission at the Monterey of Vizcaino's discovery. In 1770, this cherished dream

FIG. 58.   THE CARMEL MISSION.

was realised, and the Mission de San Carlos de Monterey was established on the 3rd of June of that year, ' being the holy day of Pentecost,' as the Father expresses it. About the end of the year 1771, the mission was moved to Carmelo Valley, some five miles from the Bay of Monterey, and called the Mission San Carlos de Carmelo. This was done by order of His Excellency the Marquis de Croix ; and here, on the banks of the Carmelo River, still stands the old stone church then erected, beneath whose sanctuary repose the remains of Father Serra and three of his co-workers, including Father Crispi, his trusted friend and adviser (Fig. 58).   The presidio, or military establishment, still remained at Monterey.   In its inclosure was the chapel, which is the site of the present Catholic church ; while on the hill overlooking the bay was erected a rude fort, the remains of which are still discernible." We also noted with interest the building shown in Fig. 59, which was the first capitol of California.

After this trip, we went on a most wonderful excursion known as "the eighteen mile drive." After going for some distance through the gently rolling fields, strewn with wild flowers, we reached the pine forests, where the pale yellow moss waves from the branches of the trees; then we climbed a hill, and through a gap in the branches gazed far out across the ocean. Thus for the first time we saw the great Pacific, and as we reached the summit of the hill, and before us the vast expanse of the blue waves rolled to the bluer sky, we wondered what must have been in the mind of the "eagle-eyed" discoverer when he and his followers stood in wonder "silent upon a peak in Darien."

FIG. 59. CALTON HALL.

Many of us went down on to the shore, to pick up the curiously marked round pebbles, and to watch the long slow waves foam gently up the beach. Soon we came to the far-famed cypress trees (Fig. 60), which writhe and twist their gaunt limbs along a stretch of wild rocky coast. It has been suggested that these strange uncanny trees sprang from the seeds of the cedars of Lebanon brought by the first missionaries, but many of them give evidence that they are even more ancient than the earliest settlers. Distorted in every conceivable shape, and hung with pale moss, they hardly seem like trees, but rather wild creatures of the waves hung with dripping seaweed and struggling to escape from some invisible chain. A little further on, the seal rocks came into view, covered with waddling

N

flapping beasts, and surrounded by others lithe and graceful in the water, while on the top of the rock perched flocks of ugly cormorants.

The next afternoon a party of five started on a horseback ride over the fine roads. The weather was such as New Yorkers enjoy about the first of June, or, in other words, perfection, and as we " loped" along beyond the town, gazing out over the curve of the beautiful bay, life seemed decidedly worth living. We passed the marble statue of Father Junipero Serra (Fig. 61), represented in the act of stepping from his boat, and located on the spot where he first landed ; and we were constantly passing lovely little houses, many with hedges of calla lilies in full bloom. We next descended

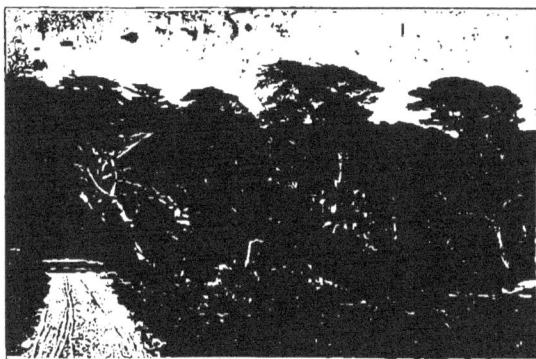

FIG. 60. GROVE OF CYPRUS TREES.

to the shore, and raced along where the bright waves rippled up to the horses' feet, while the fresh salt breezes blew over us ; then we galloped back to the hotel through the avenues of arching trees. Like many other places of less note, Monterey has its " 400." In this instance, however, they are Chinese, although it may be said they are just as exclusive as their more famed brethern of New York City, and live by themselves in one quarter of the town. After attending a sacred concert on Sunday evening, the party retired to their rooms, and the next morning we left with the greatest regret this charming Hotel del Monte and the beautiful Monterey.

The journey beyond was delightful, passing through the celebrated Santa Clara Valley, both sides of the track being equally beautiful, our route lying in front of the Santa Cruz range, some miles away, and the tops of which were frequently crowned with snow. At one point of the journey we had a view of the celebrated Lick Observatory, resting on one of the lower peaks of Mount Hamilton. Later, two of the party went there, and had the usual experience in their ascent to it, viz., a terribly hard ride, a poor meal, and an extortionate charge for transportation, but they saw the big telescope, which the rest of us fortunately were willing to take for granted. The journey was pleasantly broken by a stop at Palo Alto, for the purpose of visiting the Leland Stanford Junior University, under the guidance of one of its most distinguished professors, Professor Horace

FIG. 61. MONUMENT OF FATHER JUNIPERO SERRA.

Gale, who is a member of the Society of Mechanical Engineers. In the construction of these beautiful buildings, and the endowment of this great University, Mr. and Mrs. Leland Stanford desired to erect to the memory of their son, Leland Stanford, Jun., who died in Florence, Italy, in 1884, a noble and enduring monument.

The endowment consists of three tracts of land, to be held for ever inalienable, the rents and profits to be used for the maintenance of the University. This grant does not take effect until after the death of the donors. The trustees will then enter upon their charge, and perform the duties of managing the lands. They must either farm or lease them, but cannot sell any part of them. The total acreage of land endowed is about 85,000, consisting of 8400 acres at Palo Alto, Santa Clara County, Cal.,

where the University is located.   The beautiful summer residence of Mr.
and Mrs. Stanford forms a picturesque part of this tract.   The lawns and
gardens surrounding this mansion occupy about 400 acres.   In this area is
also located the world's famous trotting farm of 55,000 acres, known as the
Vina Estate, in Tehama County ; 4000 acres are planted in vines (the
largest vineyard in the world) ; 22,000 acres, known as the Gridley Ranch,
in Butte County, are devoted mainly to the raising of wheat.   The value
of the entire endowment is estimated at 20,000,000 dols.   It has been subsequently announced that the grantors have made additional provisions in
their wills, so that the University will have ample endowment for all time.

The original architectural study for the University was the San
Antonio Mission, shown in Fig. 62, and the result is an attractive structure

FIG. 62.   THE SAN ANTONIO MISSION.

of yellow buff sandstone with the foothills for a background.   Various
illustrations of this University were furnished by Professor Buchanan, and
the writers are indebted for the statements about the Institution to Professor
Gale.   The twelve buildings are connected by corridors, the main one
appearing in Fig. 64.   The tuition is free, and although but recently
opened, there were over 500 students in attendance, 100 of whom are
young ladies, and the indications were that before the end of the year there
would be 1000.   Board is fixed at 20 dols. per month, and there is every
inducement to a deserving student. even though possessed of but moderate
means.   It is the ultimate intention to inclose the present quadrangle by
another side two stories high, whose arcade will face outward, and to
provide a grand Romanesque arch for the entrance.

Passing through the arch now used as an entrance, the visitor comes

FIG. 63. QUADRANGLE, LELAND STANFORD UNIVERSITY.

FIG. 64. MAIN CORRIDOR, LELAND STANFORD UNIVERSITY.

upon a smooth asphalt court, with beautiful flower beds at convenient points as shown in Fig. 63. Spanish titles for the various buildings prevail to a great extent, and in fact, as this country not very long since was Mexican, the traveller finds Spanish names everywhere. Encina Hall (Fig. 65), so named from the Spanish, meaning live oak, occupies a space 312 ft. long by 150 ft. wide, and will accommodate 300 boys. It is a massive structure of five stories, and built of yellow sandstone similar to that used in the quadrangle. A portico extends the entire length of the building, at the centre of which is a broad flight of steps leading up to the main entrance. The interior is divided into halls and rooms of convenient size, each apartment being furnished with electric light, steam, hot and cold water. The woodwork is of polished oak and southern pine. The several floors are soon to be connected with elevators, and this will complete one of the most perfectly equipped dormitories in the world. A concrete walk leading from the dormitory connects with the quadrangle, and also continues to Roble Hall, which is the dormitory for the young ladies.

The Museum (Fig. 66), erected to the memory of Leland Stanford, Jun., is situated about five hundred yards north from the main entrance to the University. It is an imposing structure of concrete, 368 ft. front, and with a depth and altitude of 186 ft. and 58 ft. respectively. The entrance, consisting of a partly inclosed portico, supported by immense pillars, crowned with an entablature of Ionic architecture, is approached by a broad flight of marble steps, and is crowned by four statues representing Herodotus, Plato, Plutarch, and Aristotle. The interior of the Museum, when finished and furnished, will be well worth a visit to Palo Alto. The main vestibule is lined with Italian marble, and lighted by an immense skylight from a dome 60 ft. by 50 ft., the largest self-supporting concrete dome in the world. At the right of the vestibule is the Egyptian Gallery, named from the style in which it is finished ; on the left is the Grecian Gallery. Collectors are now in the famous Parisian and Italian galleries making a collection for the great Art Gallery of the western coast. The corresponding rooms on the second floor will contain curiosities of every description. In the main vestibule is to stand a statuary group of the family. This statue is nearing completion in Italy.

At the end of a great cypress avenue leading from the University stands the Mausoleum. (Fig. 67.) It is in the form of a Greek temple, guarded by sphinxes on either side of the entrance. This sepulchre, constructed of white marble, provides a magnificent resting place for the dead.

FIG. 65. ENCINA HALL.

FIG. 66. THE LELAND STANFORD MUSEUM.

FIG. 67. THE LELAND STANFORD MAUSOLEUM.

In a sunny spot near the Mausoleum, is a collection of rare and curious cactus plants, most of which are native of the deserts of Arizona.   Their sides gleam with sharp spears, and they look like giant sentinels guarding the approach to the sepulchre.   They are not objects of particular beauty, but as curiosities they do not fail to catch the eye of the visitor.

We next were taken through the machine shop 178 ft. by 50 ft., and adjacent to it is the Power House seen in Fig. 68.   The latter contains four 100 horse-power boilers, and the chimney is 30 ft. in diameter and 80 ft. high.

FIG. 68.   POWER HOUSE, LELAND STANFORD UNIVERSITY.

This is the first year of this University, and of course the shop is but partially fitted up, but whatever has been done has been well done, and the instruction in shop work promises to be of the most thorough and practical kind, just what those who know Professor Gale would expect from a depart-ment under his administration.   The Library was a very pleasant room, and the chapel was tastefully decorated, and extremely attractive.   The party were greatly pleased with their visit, and were taken by a lovely road to the celebrated stock farm of Mr. Stanford, and treated to a view of some of the fastest trotting stallions in the world, as for instance Palo Alto, who trotted a mile in 2 min. 8¾ sec. ; he has since died.   Sunol and Arion were also bred and trained here. The care the horses receive seems to be even greater than that bestowed on children, and certainly they were a most attractive sight, their coats shining as though varnished, and their feet carefully washed, while the symmetry of their form and their beautiful clear eyes, fixed the admiring glances of us all. But leaving these pleasant scenes, where to linger would have been a delight, we were whisked away through a constantly changing and most picturesque panorama amid vines and fruit trees, past attractive brooks and luxuriant groves, until about noon we reached the city for which we had started so long before, and were landed safe, though hungry and dusty, at the Palace Hotel in San Francisco.

The popular impression regarding San Francisco, entertained by those who have never been there, is that the city dates its origin from modern times. In point of fact it is one of the oldest cities of the United States, and was settled under the name of Yerba Buena. The founder was the celebrated Junipero Serra, whose monument was seen at Monterey, but in 1595, an explorer was wrecked in the Bay of San Francisco, and having thus landed in spite of himself, proceeded to explore the adjacent country. Mexico ceded California to the United States in 1848, and with the discovery of gold in the State, a new era dawned upon the city; the name was changed in 1847 to San Francisco, and there is a painting of the place as it looked then, still in existence.

The growth of the city under our Government was most rapid, and despite earthquakes and fires, by 1890 it had become the ninth city as to population in the United States. There was great lawlessness in those earlier times, for bands of ruffians and adventurers from all parts of the world flocked to the Golden State. Matters finally became so desperate that the better class of citizens organised a "vigilance committee," and a few sudden deaths occurring among the lawless element, and several public executions from the windows of the rooms occupied by this vigilance committee, caused a general stampede of the gamblers and assassins. With their exit there came quiet times and a healthy growth. The energy of the citizens sought constantly for new outlets of trade, and San Francisco became the foremost city on the Pacific Coast, adding new fame to the great country of which she is to-day a prominent and attractive city. In 1890 the city extended all over the sandhills and along the coast, presenting a most delightful picture. One of the most prominent and attractive buildings is the new City Hall, not quite completed, although it has been over fifteen years under construction, and nearly 4,000,000 dols. have been expended on it.

o

The completed structure is shown in Fig. 69, and is a building any city in the world may be proud of. The first impression of San Francisco is, that the city is very extensive, that the buildings are very high, and that the noise of the electric cars, produced by the incessant clanging of their warning bells is something terrific, but this last one gets accustomed to, and finally almost fails to notice.

The Palace Hotel were we stopped is, in the writers' opinion, one of the finest and worst managed places of its kind in the United States. It is modelled somewhat on the style of the Grand Hotel in Paris, being built around a courtyard closed at the top by a glass roof, and the hotel is

FIG. 69. THE CITY HALL, SAN FRANCISCO.

arranged with galleries on the four sides. Properly managed, it would be one of the attractions of San Francisco; as it is managed, it is any thing but an attraction. The food is excellent and properly cooked, but the service would have had such an effect on the Prophet Moses in twenty-four hours, as to cause him to forfeit for ever the title of the "meekest man." After trying every method in vain to obtain even notice—not attention— the sterner of the writers resolved to see what personal appeal at head-quarters would do, and he went to the office, getting no satisfaction, his sunny spirit became clouded, and he advised the proprietor to get a tailor's dummy, and put on it a dress suit, mount the effigy on wheels, and move it around in the dining room, thus saving the wages of his head waiter. But if the hotel service was bad, the hospitality of the people made amends for it; they were glad to see us, and showed it in every manner possible, even to arranging to have any of the engineers and their ladies carried free over all the cable lines in the city.

The convention was a most successful one, and the papers read and discussed were full of interest, but it is not with them this story has to do, so we will pass at once to the excursions planned for our entertainment. After being received and welcomed in a very appropriate speech by the mayor, the party were invited to visit the various systems of cable railways. San Francisco is built on a series of hills, some of them very steep, and cable lines extend in every direction. We had the advantage of making this trip under the direction of the original projector of these lines, Mr. A. S. Hallidie, whose untiring efforts contributed greatly to our pleasure and instruction. The system was put in operation about 20 years ago

FIG. 70.   THE GOLDEN GATE.

and the speed was but four miles an hour. There are now upwards of 20 miles of cable lines in San Francisco, and their effect on property has been wonderful. Lots costing originally 600 dols. each, are to day worth several thousands. The city commenced to grow in those directions where these lines were laid, and nearly 2,000,000 dols. is invested in the cable system.

Another excursion which was greatly enjoyed, was a trip taken in a large ocean tug down to the Golden Gate, Fig. 70, as the entrance to the harbour is called, and in going there the party passed along the water front. The view of the great city as it lay spread out over the slope of the hills was most pleasing. As the fast sea-tug sped along, and we drew nearer and nearer to the opening between the rocky cliffs which marks the entrance to the harbour, we began to realise that the water was not in the glassy and calm condition that the Pacific should preserve, and that the tug was obeying the laws of gravity to an extent that threatened to be alarming. Many of the party grew solemn, and reflected on the uncertainty of human life.   But we still sped towards the open ocean, and soon the long blue waves began to swing the tug, although she was a large boat, on their crests ; they were not angry and turbulent, but were cruelly deliberate, long slow waves, taking plenty of time for their full effect, tranquil as oil, with a steady rise and deadly sweep down, with which every landsman is so sadly familiar.   One of the managers of the party, who was seated on the upper deck enjoying the sail, noticed an unusual quiet on the lower deck, quite different from the hilarity which had prevailed at the start, and went below to inspect and ascertain the cause.   One look around was sufficient, the silence was far more eloquent than language, and the tug, which had been assuming the various vertical angles incident to a condition of unstable equilibrium, was turned through a horizontal angle of 180 deg., just in time to prevent a crisis, and we headed for Mare Island, which was soon reached after a most delightful sail.   The Government Navy Yard is located here, and it is well worth a visit, to say nothing of the pleasure of going there. Some of the White Squadron were here, and after being welcomed by the commandant, the party inspected them, and a few, the writers among them, enjoyed a breakfast on the cruiser " Boston."

San Francisco has a water company that could supply enough of this fluid to saturate the rail-stock of Wall-street, and after this no one will doubt its capacity.   The party were invited to inspect the Crystal Springs Dam, and accepted with much pleasure.   The Southern Pacific Railway took them by special train to San Mateo, and thence they were conveyed in carriages some ten miles further.   The beautiful residences in the centres of parks, under a system of complete landscape gardening, were the remark of all, and those fortunate enough to have visited England in 1889, with this Society, were frequently reminded of some of England's finest mansions. The profusion of exquisite flowers growing everywhere out of doors astonished the visitors, and lent magnificent colour to the scene.   The dam itself was a noble structure, the wall being in places 145 ft. in height.   It was stated that 250,000 barrels of Portland cement were used in its construction.   This delightful day was appropriately closed by an open air lunch spread beneath enormous trees, and where a fluid more exhilarating than spring valley water, flowed in abundance.

One more trip in San Francisco remains to be chronicled, and that was incident to an invitation from Adolph Sutro to breakfast at his residence and inspect the Cliff House and Sutro Heights. No one visiting San Francisco should fail to visit these two places, for they are reached by a short trip in the cable and steam cars, of about half an hour. Our host met us at the entrance to Sutro Park, and conducted us through it.

The work Mr. Sutro has done here is a shining example of pure philanthropy. A barren cliff surrounded by sandhills, and facing the open ocean, was all there was to be seen in this place ten years ago. Mr. Sutro has laid out a beautiful park with handsome shrubbery, fine lawns, and gravelled walks. He has filled the place with rare and beautiful plants and trees, adorned it with statues, and thrown it open to the public for their pleasure ground. The cool breezes from the ocean render this a delightful spot to all, and the poorest classes, as well as the rich, find pleasure and health from Mr. Sutro's liberality.

FIG. 71. THE CLIFF HOUSE, SAN FRANCISCO.

A few animals are to be seen, such as monkeys, bears, and deer, that lend interest to the scene. The cliffs are, however, the great object of interest; they are on one side of the shore of the Golden Gate, and are about 200 ft. above the ocean.

The Cliff House, shown in Fig. 71, is a restaurant, a place for basket lunches and an outlook into the open Pacific. The shore below is filled with happy children playing in the sand, or wading in the water. But the great attractions at the Cliff House are the Seal Rocks, shown in Figs. 72 and 73. As these animals are protected by law, they have multiplied rapidly, and are a never failing source of entertainment. It is impossible to imagine any animal having a better time than the seals, and when swimming in the water they are a most pleasing sight. Their heads with their large and

beautiful eyes, have a wonderful appearance, but when they get on the
rocks, and waddle with a heavy lumbering motion, wriggling their huge
satiny bodies along the earth, they resemble huge slugs.   The waves are
constantly breaking against these rocks, and one of the favourite amuse-
ments of the seal is to hold on close to the edge and let the waves wash him
off.   The old bulls frequently indulge in a fight, and butt each other or

FIG. 72.   SEAL ROCKS, SAN FRANCISCO.

FIG. 73.   SEA LIONS, SAN FRANCISCO.

strike with their flippers, uttering the most tremendous roars of rage.
Their size is very great, and it was stated they will consume from 40 lb. to
60 lb. of fish each per day.

Mr. Sutro's philanthropy does not stop with the park he has laid out.
At the time of our visit, he was completing an enormous marine aquarium,
having excavated, at a great expense, a large plot containing many acres at

the base of the cliff, and in the centre he has left a part of the original rock, which will be an island. Adjacent to this aquarium are very large swimming baths provided with an arrangement of steam pipes by which the water may be heated to any desired temperature, the floors of them being cement. This is to be free to the public. The method of supplying the sea water shows not only that Mr. Sutro is an engineer, but that his experience with the celebrated Sutro Tunnel is still in his mind. The sea water comes through a tunnel connected with a large storage basin. The latter is so located that at high tide the water dashes over its sea face, then by means of gates it can be fed into the aquarium or baths as desired. Any sand in suspension is caught in transit and returned to the ocean. Mr. Sutro's pastime is to inspect all this work daily, and oftener even, and this growing monument to his munificence and philanthropy must be a constant source of delight.

After leaving this hospitable mansion, the party were driven through Golden Gate Park for the city of San Francisco, having seen what Mr. Sutro has done, and quietly learned its lessons. He is busily engaged in making a most attractive park out of the adjacent sandhills. About 1013 acres are contained in this tract, and fine roads and promenades are to be seen in all directions. Several beautiful statues have been erected, one of the finest being that to President Garfield. The great attraction of this park, is, however, the conservatory, constructed of glass and iron, and 250 ft. long by 60 ft. in width, the dome rising to a height of 58 ft. It contains many exquisite varieties of flowers and a fountain of unique design — viz., a swan surrounded by magnolias upholding a bowl in which is a mermaid holding on her shoulder a sea shell, in which the fountain plays.

The suburbs of San Francisco were the next points visited ; they well repaid the excursion. No visitor should omit seeing Berkeley and Oakland just across the bay, both easily reached by ferryboat. These places are principally occupied as residences for the San Francisco business men, and are full of beautiful drives bordered by many magnificent houses. San Francisco has also several fine clubs, one of the principal being the Pacific Union ; near by it is the Cosmos Club, one of the most home-like places imaginable. The Bohemian is also celebrated, taking in, as its name implies, journalists and actors for the most part, although comprising in its members some of the leading men of the city. That evening there occurred one of those hospitable receptions which serve to make life brighter.

This party, like all of Messrs. Raymond and Whitcomb's, was under the personal charge of a conductor, to whom we looked for everything, even our daily bread. Now these gentlemen, with that great knowledge of detail, which has made their parties so delightful, have selected their conductors with great skill, and our special one was a man whose nature was as gentle and kind as that of a woman, and yet when necessary, he could be as firm as a rock ; he was, in fact, a true gentleman, and a man of cultivation. His conduct of the party had been so happy that by one of those spontaneous movements it was decided to present him with a slight

FIG. 74.　C. A. COOKE.

token of our respect and esteem, and this was done in the hotel parlour by one of our party, who made a most happy little speech, and gave him a very pretty locket with the Society monogram in diamonds on it. Our good friend Mr. Cooke was completely overcome with pleasure, and expressed his embarrassment in a few words, blushing like a school-girl, to the great delight of all the ladies, who could hardly imagine this large 6 ft. 2 in. man showing, what in them would have been considered a weakness.

The writer is glad to chronicle this little episode and to suggest to others not only to go to California under Raymond and Whitcomb's auspices, but under Mr. Cooke's personal charge. It is really the perfection of travelling ; the sole trouble is that it spoils one for ordinary methods. The passenger under these auspices simply takes care of himself.

The entire party are registered in advance of their arrival at any place; just before leaving the train each traveller is handed a ticket bearing the name of his hotel and the number of the room assigned to him; he just takes the elevator and goes to it. Frequently he finds his baggage has been sent in advance, and is awaiting him in the room, or, if not, it is sure to be there within half-an-hour. Even hand baggage is taken charge of, and as the hotels are invariably the best in the place, and he is taken there from the station, it is impossible to grumble with any show of reason. The writers are not advertising these gentlemen, but simply stating the facts as verified by their personal experience.

No visit to San Francisco is ever considered complete until a trip has been taken through Chinatown, and receiving an invitation from a city official to accompany a party of Boston aldermen, guests of the city, one of the writers saw this portion of San Francisco in its entirety and to its dregs. It is all very well for eastern men, and for foreign nations, to talk about the inconsistency of Americans in proclaiming America as a land for the oppressed of all nations, and then labouring to exclude the Mongolians. Let any one who entertains these sentiments—because such views are purely sentimental—if he is a man of intelligence, make one careful visit to the Chinese quarter in San Francisco, and he will be confronted with a problem which cannot be solved by theory, but will require the most careful study of a real practical nature. It is not that Chinese are more vicious than other nations; to their credit and to our shame, it may be said, they can claim a favourable comparison with a far higher civilisation. But they establish themselves in a strictly exclusive community, whose whole aim is hostile to the principles of our government; they live in a way that is demoralising and degrading; they are not amenable to any influences tending to improve them, and, what is worse than all, they never assimilate with the people around them so as to be brought in contact in time with anything which may tend to elevate them. Living as they do, they are brought into direct competition with our labouring classes, and are enabled to affect seriously the source of livelihood of our citizens. They add nothing to the wealth of the land, but in time return to China, taking with them the money they have accumulated. This is neither the time nor place to discuss the Chinese question, but with these thoughts in view, which are full of suggestiveness, the reader is asked to go to Chinatown, and the distance is so short that in five minutes' walk from the best parts of San Francisco the visitor finds himself in the heart of this foreign quarter.

P

Fascinating place ! where one is suddenly transported from the western city with all its high civilisation, to the very heart of China, where the wearers of the pigtail swarm on every side in garments of gorgeous hues, where the strongest odours fill the nostrils and the strangest sights greet the eye. As for the shops, with their gilded carvings and embroidered stuffs and tiny images, and the sweet perfume of the sandal-wood, they were simply enticing, and our ladies spent a large part of their time there, trying on one loose robe after another, admiring the delicate china, and attempting conversation with the affable little Celestials who patiently followed them about. One round-faced boy was particularly untiring, and showed us carved ivory and lacquered trays without end. As he spoke very little English we communicated principally by signs and nods, he grinning with delight when he caught our ideas. We bought a number of trifles, which he carefully packed in straw, and on our asking if he would be sure to send them in time, he replied with a smile of pride—

" Yes, yes ; five between six."

One of our party who was an enthusiastic photographer was much disappointed in Chinatown, for the people ran from him as if from a plague, and though he spent a whole afternoon there he was utterly unsuccessful. Taught by experience, however, he returned the next day and adopted the following ingenious tactics. He would stroll along the street in a most nonchalant manner, with his camera open and set, staring at the buildings and behaving rather strangely. Gradually a small crowd, principally composed of children, would follow him at what they fondly supposed to be a safe distance. When the pattering footsteps assured the fox that the geese were close at hand he would quickly shake off his abstraction, wheel about, and snap them as they scattered. Some of the results were excellent photographs, and others were laughable views of flying pigtails. The streets are thronged at all hours of the day and night with Chinese, both men and women, boys and girls, for in the space of four or five city blocks are packed some 30,000 of these people. Chinese shops line both sides of the streets, and are extremely interesting to the visitor. The drug stores are a study, all the contents being brought from China. Snake skins, powdered locusts, together with many herbs and drugs, are on the shelves. Dried fish, edible birds' nests, fruits, nuts, &c., are to be seen in the grocery shops. What the butchers' shops contained, not even a Solomon could have decided, for there is no nation more adept in the art of concealment, and they carry it to the highest point of perfection in their viands. The joss houses are to be seen also, and a visit to them is quite interesting. The walls of the entrance are covered with what seems to be the labels to fire-crackers, but on inquiry we were told they were the names of subscribers to a collection. If this was true, the Chinese in America are liberal to their gods. On ascending one flight of stairs we were met by a priest, who sold us bundles of fragrant wood to be burned as a prayer or offering. There were embroideries on the walls of dragons and other monsters in heavy gold work. There were three hideous idols placed behind tables, on which were offerings apparently of fruit and food, but evidently the Chinaman has found this expensive, so, with true economy, he has substituted for the real article an imitation in coloured wax or some other plastic material. The ceremony begins with sounding a gong to wake the joss, who is always in a somnolent state, then little tapers are burned before him ; the priest prostrates himself, and, for all the visitor knows, may be beseeching the joss to remove the intruder to Sheol at the earliest opportunity. There was a large spear in the joss house, which the Chinese priest declared very heavy, and only to be handled by the joss, but we lifted it with but slight effort. The next place of interest visited was the Chinese restaurant. It was clean and attractively furnished, and remembering St. Paul's injunction, " Eat those things that are set before you, asking no questions for conscience' sake," we followed it. The tea was very fine, and the sweetmeats were delicious. The cigars, too, were satisfactory.

If Chinatown was enticing in the daytime, when the pitiless sunlight showed all the dirt in the streets, how doubly fascinating was it at night, when in the uncertain light of the swinging paper-lanterns it seemed like a strange scene from a fairy tale ! Gazing down the dimly lighted street, a bright spot showed a goldsmith sitting cross-legged before a jet of flame, his yellow face bending over his work. It was interesting, too, to stand near some butcher's shop and watch the people going in to buy the strange-looking wares, or passing by in the square of light before the door, the women in their loose, graceful robes, some of green or scarlet silk, embroidered with gold, and the toddling children clothed in four or five colours at once, with pink and blue silk braided in their shiny cues.

"But," says the reader, "all this is interesting, and nothing appears to justify your charges against this people." No, my friend, all this is above ground. Now descend at night this flight of steps in a dark alley, and enter into a narrow passage some 10 ft. below the ground. You would probably imagine yourself in a mine, passing along one of its gangways.

The guide opens a door, after having taken the precaution to tell you to puff well on your cigar. You notice there are similar doors along this passage at intervals. You see seated on a stool a figure smoking, and note around this den, which is about 6 ft. square, that there are three tiers of rough bunks, each containing a sleeping figure. You are indeed in " underground China." In this room are twelve Chinamen, with absolutely no ventilation or light, except from a small lamp, who are consuming whatever oxygen the lamp has left, by their tobacco or opium, as the case may be. There are thousands literally living just this way. A little larger room may contain more men, that is all. The idea is to give each man just enough room to lie in, say 6 ft. by about 2 ft. Even swine have more room. The food is cooked over a common fire, and consists mostly of rice or refuse, perhaps both. The sanitary condition is something appalling, and why they don't die by the score no one can say. Any other nationality would be extinct from this cause alone. This visit was carried on for a considerable time, with occasional retreats to the surface for air ; the writer determined to see for himself and not to accept any one's statement. A few years ago this trip would have been attended with great personal risk, and ended abruptly with the disappearance of the seeker ; now, however, it is not dangerous, unless one encounters a drunken Chinaman and has a row with him. In one den we saw living in a place just large enough to hold three persons standing, an insane blind Chinese woman, with a parrot, a cat, and a dog. It is impossible to depict these scenes any further in a history like this. Let the reader's imagination be stimulated with the statement that there are worse scenes than these described, as to their filth and squalor, and we will pass at once to a more cheerful Chinese subject, viz., the Chinese theatre.

The entrance to this place is through an underground passage, with small dressing-rooms leading off it, and the doors were open so that we were allowed to see the actors "making up" for their various parts. Here the similarity of features between men and women was an advantage for all the female characters are personated by men, and with a little rouge and a judicious use of paint, the deception is perfect. After a few words with these people, we ascended a short ladder, and found ourselves at the back of the stage, and in front of upwards of 1000 people. The whole lower floor or pit was packed with Chinamen, all standing, while the women and children were in the gallery. A place was made for us at the right hand and alongside the orchestra, who never cease playing. The music may appeal to the Chinese ear, but it seemed to us the most dreary sound imaginable. There is no scenery, and the action took place mostly on the left side of the stage. We were so large a crowd that we encroached very much on the space allotted to the actors, who were obliged to make their entrances and exits through our number, but this did not seem to bother them, nor did our presence and conversation. The play went on, and even the terrible fight where the villain stabbed the heroine in the face, and her blood literally ran upon the stage (by some Chinese trick), occurred within reach of our arm. The coquetry of the female impersonators was quite worthy of the original type, and their airs and graces were as natural as possible. What all the play was about we did not know, but the Chinese audience were very much interested, and showed as much excitement as a Chinaman ever allows himself. The women and children were more moved, and smiled and sobbed according to circumstances. The actors' costumes were very elaborate, especially the dresses of the female characters in a wedding scene. In case of a panic or a fire, escape would have been impossible for most present, and realising this we did not stop to see the conclusion, but slowly threaded the dark passage and found ourselves in the alley-way. We then visited gambling houses, and saw the most imperturbable countenances when luck ran against them ; not even a wink from the loser of more than a week's wages on a single jack pot, for they play American poker with as much grace and boldness as an old Californian native. Nor did the winner show any signs of exultation. The game went on in perfect silence, each man indicating the amount he wished to bet by showing it, and no one uttering a sound. On our offering consolation to the loser with the remark, " Better luckee next time," he replied without the slightest emotion, " Badee luckee, three weekee washee gone hellee."

The last evening in San Francisco was marked by a delightful reception, where we were privileged to listen to a most interesting paper by a lady whose mother was one of the early settlers in California. In this the daughter described most graphically those exciting days when the gold fever drew eager crowds from the east into the wild forests, where some made a fortune in a day, and others lost money, health, and even life in a hopeless unrewarded search. At the close of the evening each lady was presented with a large bouquet of rare sweet peas of most beautiful and unusual colours, which had been sent by a gentleman who had given much of his time to their cultivation.

Before starting on the Yosemite trip the tourist should most certainly go to the office of the Stage Company, opposite the Palace Hotel, and meet Mr. Samuel Miller, the agent for this company. He will not only form the acquaintance of a most amiable and genial gentleman, but will get valuable suggestions of his journey and be able to see views of the glories to come. Half an hour spent there will be well employed; the writer speaks from personal experience. Having completed our arrangements, we started the next afternoon to realise all the bright dreams of our boyhood, for to us

FIG. 75. SAN FRANCISCO TO THE YOSEMITE.

FIG. 76. PLAN OF THE YOSEMITE VALLEY.

the Yosemite had always been a vision of the future, something we had fondly hoped might become a reality, and we may say at the outset that we were not disappointed in any particular. As we left San Francisco by the Oakland ferry-boat a sea-fog swept around the great headlands, and by the gleam of the setting sun made the entrance in reality a "Golden Gate." A map showing the location of the park in its relation to San Francisco is given in Fig. 75, while the valley itself and the various trails to different points of interest, appear in Fig. 76.

The next morning found us at Raymond, where we had to leave our pleasant conductor, Mr. Cooke, for a season, and commit ourselves to the guidance of the Yosemite Stage Company. But our trust in this company was not misplaced; they have systematised the transportation in a thorough manner, and the journey, which of necessity is somewhat fatiguing, is made as easy as possible. A large amount of money is spent annually upon the roads, and the stages are roomy and comfortable. The drivers are careful men and full of interesting stories, which they will unload on the slightest provocation. The horses are most excellent; they are strong and are well cared for, and start with a run, showing they are not overworked. Relay stations are placed at seven to eight miles apart, and it was stated that over 400 horses are employed in this service. So then the reader may imagine the party after a good breakfast at Raymond, seated in their stage, and ready for departure.

The writers approach this part of their narrative with great misgivings, for a proper description of the glories of the Yosemite should be written by a Ruskin and be illustrated by a Turner. Thus only would full justice be done to the magnificence of the various spots we visited, and the pictures that were presented to us. But as this is impossible, the reader must be contented with the plain statement of facts, and let his imagination supplement the accompanying illustrations. The distance to the Wawonah is 36 miles, and the time necessary is eleven hours, with one hour's stop for lunch, and with relay stations every seven miles. The day was warm and the dust was prevalent, so that occasionally at the relay stations we had to be excavated, for we were gradually becoming like the remains found in Pompeii. The road also was quite steep in places; and our stage took vertical angles of more or less inclination as we proceeded. There was no special beauty in the rolling ground over which we were passing, except the wild flowers. It is impossible to describe them, for they were of every colour of the rainbow, and of an innumerable variety of shapes. The bushes around us were white with perfumed bloom, and in the grass were a thousand exquisite flowers, so tiny that they seem made for fairies. We gathered quantities of them, and were constantly jumping from the coaches for this purpose, but the lovely wild things faded almost as soon as they were plucked, like the scented rushes of Alice's dream. We lunched at Grant's, where we were separately dusted before we entered the house, and after an excellent dinner and a short rest we proceeded on our journey. Now we climbed steadily, and soon we were in the pines. All around us the great trunks rose into the air, the bark of some beautifully marked and seamed, while high above our heads the feathery green waved against the deep blue sky.   Once, at a turn in the road, the mountains parted and we looked through the clustering peaks far out on the plain below, which was gleaming in the amber sunset light.   Soon the sky was aflame with gold and orange, and full of drifting pink clouds, and as we plunged deeper into the mighty forest, still the bright glow shone beneath the trees.   Then in the space above us gleamed the stars, and the long shadows crept across the ground, making the great trees even more awe-inspiring.

We reached the first night's station at Wawonah, and stopped at the hotel.   The Palace Hotel at San Francisco was cited as one of the worst managed hotels in the writers' experience, but Wawonah is supreme in their mind as surpassing it.   And what is most aggravating is the fact that the proprietors apparently mean to do what is right, but are utterly without any system in their management.   As there are three heads to the establishment, it is utterly impossible to fix any responsibility.   If you complain of anything it is always the brother who is absent that is guilty, in accordance with the French proverb, "The absent are always wrong."   One of the writers desiring rooms for his ladies, was told that one of the brothers had actually slept in a bath tub the night before, and he ought to be glad with any accommodation.   He replied, there was no accounting for tastes, and if the gentleman was a descendant of Diogenes that was his misfortune, and added he had remarked that the brother was round-shouldered and was glad to know the reason.   It is best to keep your temper, however, for you are absolutely in their hands, and as they control the majority of the stock on the stage routes, there is no escape, you have got to go their way and stop at their house.   The food is good and the rooms clean and comfortable, so all that is needed is one head, and that one belonging to a systematic man.   The situation of the hotel is very picturesque, and the studio of an artist who lives opposite the house is one of the attractions of the place. A visit to it is most restful, and the gentleman himself is one it is a pleasure to meet.

One of the enjoyments of the hotel is, after your own party is located, to sit on the piazzas and watch the struggles of the new arrivals, and if you have kept your temper and succeeded in obtaining accommodation, it is most amusing to see the outbreaks of others, knowing in advance they will only fare worse for indulging in their natural inclination.

We left Wawonah at about seven the next day for a 26-mile ride to

Stoneman House, and were due there at 2 p.m.   It was delightful to whirl
away in the fresh cool air and sparkling light of the early morning, through
the pines where the monster cones lay beside the road, where the brilliant
emerald moss clothed every dead limb, and the mistletoe hung in garlands
from the oaks.   And still we climbed until we were beside the snowbanks,
and all left the coaches to plunge our hands in it, and gather the gorgeous
red snow flowers.   Many of us walked long distances in the bracing air,
stopping to drink at the icy springs and the shallow brooks through which
the horses splashed.   At noon we reached the cliff above the Yosemite,
and as we gazed down the wonderful valley, a silence fell on all of us.

FIG. 77.   THE YOSEMITE FROM ARTIST POINT.

Such a view as this may not be described, and the reader is referred to
Fig. 77, which will convey a faint idea of it.

As we descended we heard the noise of the many cataracts, whose
echoes fill the valley with a never ceasing sound ; on every side they clashed
and sparkled over the cliffs, combing down in a dozen shining lines, or
leaping abruptly in a foaming stream.   On one side the little Ribbon Fall
poured through its dark gorge in wavering foam, making one leap of

Q

2000 ft., and another of about 1000 ft. more ; on the other, the Bridal Veil (see Fig. 78) darted like a thousand snowy arrows, falling 860 ft. at one leap, while its cold breath blew on us as we passed, and just above it swept the thin silver of the " Widow's Tears " (so called because it runs dry every six months). Behind the pale front of El Capitan, which rises to a height of 7012 ft., and is shown in Fig. 79, were masses of purple

FIG. 78. THE YOSEMITE; BRIDAL VEIL.

cloud, and the thunder echoed through the cañon, lending its sound to that of the rushing water, and seeming the fitting voice of the Great Spirit whose abiding place is the steeps, and the mountains, and the mighty waters. In the valley raced the Merced River, dimpling in the rain, which now fell in big drops.   The aromatic perfume of the woods rose in the dampness, and the odour of the flowers grew more sweet.   We heard the crash of the Great Fall of the Yosemite, and saw its white waters plunge 2548 ft. on the rocks beneath, shrouded in floating grey mists.   We reached our

destination, Stoneman's, exactly at two, for these stages leave and arrive
with the punctuality of trains, and we all agreed that we had already
been fully paid for the trouble of our journey.

Although one might be supposed to rest after this fatiguing journey,
yet such is the inspiration of the scene, that all bodily fatigue is forgotten,
and the visitor hurries forward to see any new glories he may have over-

FIG. 79. THE YOSEMITE; EL CAPITAN.

looked. That same afternoon we went across the river, and walked about
three miles to the foot of the Yosemite Fall. On arriving, we climbed to
the foot of the upper fall, amid spray so thick as to be almost blinding,
and requiring the protection of a good waterproof overcoat. The scene
well repaid us, and all fatigue was lost sight of in the glorious sight before
us. Let the reader imagine himself on the rock in front of this fall, which

is 34 ft. wide at the top and falls 1502 ft. at a leap. He may look down on a series of cascades, which fall about 600 ft., and still lower may see 400 ft. to the lower fall. Just preceding the time of our visit there had been heavy rains, and every fall was seen at its best. June is the best season to visit this place, although some prefer September. Our experience would lead us to select June, as in the later month the falls are not nearly so large owing to the low water.

FIG. 80.   CATHEDRAL SPIRES, YOSEMITE VALLEY.

We rose at six the next morning, and after breakfast started through the woods in an easy riding stage. We looked with ever-increasing wonder and delight at the great cañon around us—the twin Cathedral Spires shown in Fig. 80, which attain a height of 5934 ft., and at the peak of Sentinel Rock, 7065 ft. above the valley, shown in Fig. 81 ; near this is Sentinel Fall, which is the highest in the valley, and although quite small has a leap of 3270 ft., which it takes with great calmness, and looks like a beautiful silver thread upon the rocks at the base. The day was a perfect one, clear sunshine after rain.

The pine needle glistening with moisture shone like satin in the pale morning sunlight, but as we penetrated deeper into the woods we entered the gloom of twilight where the sun had not yet come, and we reached Mirror Lake (Fig. 82) in the morning light. Every cliff and every tree

FIG. 81. SENTINEL ROCK, YOSEMITE VALLEY.

was here reflected as in a glass. We saw the snow-capped peaks, and the blue sky, which grew momentarily brighter, and still gazing in the mirror, like Perseus in the shield of Minerva, we saw the gleaming disc of the sun rise over the cliff, while in the woods the birds burst into song, as though heralding this scene of beauty. For an instant the lovely view lasted, even one tiny white cloud glowing on the bosom of the lake, and then, as the sun's rays shot out across the water, the picture on its surface vanished, and it was but as a muddy pool.

After driving back a mile we met our guide and horses, and proceeded, Indian file, up the narrow Nevada trail. Could anything be more beautiful? Constantly opening vistas down the chasm and through the valley; now we

FIG. 82. MIRROR LAKE, YOSEMITE VALLEY.

crossed a bridge with a torrent roaring beneath, now we stopped to water our horses at a clear spring, or we skirted the edge of the cliff and looked down its sheer sides to the river, as one man said, "with only a mule's hoof between me and eternity." At length we came out on a clearing whence we commanded a view of the entire valley. Beyond us the great Nevada Fall, 617 ft. in height, whirls its gleaming rockets of foam as if out of the blue sky; half way down striking a mass of rock, it leaps high into the air in showers of spray. Then the river roars on through the valley to the Vernal Fall about a mile below, where it rolls over a precipice of 400 ft.

like a sheet of snow, the mist drifting far down the valley and covering the rocks with bright green moss. Our animals were either mules or horses, according to the weight of the rider ; the distance was not very great, being only about 4½ miles each way, but the ascent was about 1400 ft. This does not sound formidable, but it was condensed into a comparatively short distance, so that it was frequently necessary to bend back in the saddle to preserve a proper equilibrium. The gait for ascent and descent never exceeded a good walk. The living here was excellent ; the horses and mules were kind and surefooted, while the guides were nice careful men ; and, lastly, the charges were quite reasonable, and no one objected to them. Descending still lower, the clouds began to gather behind us as though desirous of furnishing the explorers of their territory with a suitable escort to the frontier ; it grew darker and darker, until there was every indication of a heavy rain. Occasional flashes of lightning could be seen, and the rumble of the thunder reverberated from cliff to cliff as though caught up and hurled from one mountain to another ; but nothing occurred beyond this, and as soon as the level was reached the entire party started on a good gallop and reached the hotel amid the pattering of the rain. "Surely," the reader will say, " you now were in a condition to rest." Perhaps so, but maybe the reader has never been in the Yosemite and felt the impulse, constantly accelerating, to keep on and see all possible points of interest. A friend of the writers' told them that he had been there a number of times, and he had only one rule, which was to take everything in the shape of a trip which the guides suggested, and as fast as it could be undertaken. We followed this advice strictly, and were only stopped by darkness from pursuing our excursions.

So we snatched a hasty dinner, and were soon seated in a comfortable stage drawn by four active horses following the road along the beautiful Merced River to the end of the cañon some nine miles distant. This stream, clear as crystal, roars alongside the road, tossing foam and spray high in the air, having in many places most beautiful rapids, and in others cascades sometimes 8 ft. to 10 ft. high. When it is added that this route runs past the foot of the Yosemite Falls on one side of the river, and in front of El Capitan, past the Bridal Veil Falls on the return trip, and that at the end of the cañon a fine view of Cascade Falls is to be had, the reader will understand what attracted the tourists. At a narrow place, with the rocks on one side and the river on the other, we had the excitement of meeting another large wagon which contained a very fussy old lady, who screamed like a fog-horn as our wheels locked theirs, and sprung from the high front seat with a surprising agility considering her years. She sheltered herself behind a tree, from which safe refuge she shrieked a succession of undesired suggestions to the gentlemen, who finally succeeded in disentangling us. Beautiful cataracts fell from the cliffs on either side, and the Bridal Veil was wreathed with a gleaming rainbow. At the end of the valley is the Cascade Fall, which is different from all the others, as the stream, severed at the summit, unites half way down in clouds of foam.

Early the next morning we started on the trail to Glacier Point, galloping down the valley on our swift little Indian ponies, and fording one swollen stream so deep that we had to hold up our feet to escape the water. Let no one miss this trip to Glacier Point if it is possible to take it, for it is a sort of general summary of all the trips together. The distance is about six miles, all but two being the ascent. It is not more fatiguing than the trip to the Vernal and Nevada Falls, being only a little longer and somewhat steeper. After pursuing the trail for a time the rider emerges from the forest and has a most enchanting view of the Yosemite Falls on the opposite side of the cañon; still rising, he is finally on a level with the top of the lower fall, and then, by gradual ascent, he reaches a point where the entire cascade connecting the upper and lower falls is visible : lastly, he has a view of the entire fall from top to bottom, having risen to the level of the top of the upper fall. But the ascent continues and the view widens in extent, till, finally reaching the summit, and looking in one direction, a range of many miles is taken into the scene, and the explorer has attained an elevation of 7201 ft.

Turning in another direction, a different landscape is shown, even more extensive and diversified. In this view are both the Vernal and Nevada Falls, the latter of which is, in the writers' opinion, the finest fall in the Yosemite Valley. In the background stretch the far-off ranges of mountains, from 40 to 60 miles distant, while the vast gorge of the cañon is at your feet. The Stoneman House on the valley level is a hencoop, and the people there look like ants. There is a small hotel at the summit, kept by a most genial proprietor, and where a very excellent meal can be obtained. At night this gentleman collects a pile of the gigantic pine cones, and, after firing them and letting them burn, producing a beautiful effect, he dumps them over the side of the cliff, their descent seeming very like a shower of brilliant stars. These fireworks are the nightly enjoyment of the guests at the Stoneman House, and all have a kindly feeling for the host of the Glacier Point Hotel.

At six on Wednesday morning we said good-bye to the Yosemite with deep regret. As we slowly climbed the ascending road, view after view of it opened before us. The sun was just rising, and while all the great peaks in the west, the Half Dome, and the Sentinel Rock, were in purple shadow, the white light shone on the pale, bald front of El Capitan. At noon we reached Wawonah, and after lunch started for the big trees. The writer can unhesitatingly recommend this drive as a sure cure for dyspepsia, and the wild leaps of our coaches would have put an Irish jaunting car or a Canadian caliche to shame. After we had been rattled like popping corn for some distance, we wisely decided to rise in the trot, which we proceeded to do by holding on the back of the seat and jumping at the jolt.

The trees were wonderful, immense, overpowering, their deep-red colour and the corrugations of their bark greatly enhancing their beauty. Towering to the height of nearly 300 ft., one-third of which is the elevation of the first limb, they present a spectacle of the greatest impressiveness. There are some 365 trees in the upper grove, but all are not of the largest size. They are estimated to be from 1000 to 4000 years old, and flourish in altitudes varying from 5000 ft. to 7000 ft. Their botanical name is something difficult, " Sequoia gigantea ; " at all events it suits them well, for the grove seems to thrive under it, and to hold up their heads with a family pride.

After passing through this grove, the appetite of the traveller is whetted for something greater, and it is not far away, for after bumping and thumping for a few miles, he descends and measures the tree known as the Grizzly Giant, and proceeds to verify the statement that it is 100 ft. in circumference, and that the " Governor " is about 76 ft. But the greatest curiosity is the " Wawona," through which the stage drives, the opening being 27 ft. deep and 10 ft. square, as shown in Fig. 83, while the tree flourishes green above one's head, maintaining as much apparent indifference to this excavation as a man would make to a piece of skin knocked off his knuckle. There are seven trees which range from 250 ft. to 272 ft. in height, and ten from 220 ft. to 250 ft.

The absence of young trees is caused by the fact that it takes the cones holding the seeds some four years to mature, and the chances against their preservation from animals and storms are very small, so that it would seem

R

to be only a question of time when this species becomes extinct.   This
15-mile drive was concluded in time for supper, and the next day we started
for Raymond.   The road lay through woods full of wild flowers, and was
cool and shady.   At the summit of a hill we saw an old man with a long
white beard and long silvery hair making his morning ablutions in a tin

FIG. 83.   THE "WAWONA"   SEQUOIA GIGANTEA.

wash-basin.   He bowed gravely to us as we passed along, and we were told
that he was the hermit of the Yosemite, who has lived in a log cabin for
30 years or more.   The mystery of this is unexplained, and while the old
man, who must be upwards of 80, is pleasant enough, yet nothing will
induce him to break away from this solitude.   The driver informed us that
he had never seen a railroad, and that he had tried in vain to induce him to
take a seat in the stage, when a few hours would put him in Raymond.

The only enjoyment he permitted himself was to occasionally go to a settlement and sell his produce. Having done this, and procured a jug of whisky, he would return to his cabin, and, in the driver's language, "enjoy a high old lonesome." That night we were in Raymond in time for supper, and, finding our familiar sleeping car and our own porter, we were soon speeding for San Francisco.

Nothing of special note marked the trip from Raymond, unless it was the transfer boat employed at the crossing of the bay near Benicia. The

FIG. 84. YOUTH AND AGE.

train was run upon the deck, and the whole transported across the straits. The boat is 425 ft. long, and 84 ft. beam. There are four railroad tracks upon its deck, each capable of holding 12 freight cars, or 48 in all, and there are two compound condensing beam engines with cylinders 65 in. in diameter, 11 ft. stroke. Each of these engines runs one paddle-wheel, these wheels being at the side of the boat as usual, except that one is ahead of the other sufficient to allow the cranks to pass, the cylinder of one engine being forward of the shafts, while the other is aft. The wheels are 30 ft. diameter and 17 ft. wide, each being handled independently by its engineer under direction of the pilot, steering being done by this means as well as by the steam steering gear.

It was full moon, and the crossing was a matter of great delight as a scene of beauty. But there was no time for the party to delay at the Golden Gate, charming as this was to us, for Alaska was to be visited, and this seemed like a trip to the North Pole to our unsophisticated minds. So the party simply got together their various impedimenta, and we were soon in our own train and with our own conductor, speeding northward along the boundary of the Pacific Ocean bound for Portland, *viâ* the Mount Shasta route, which abounds in beautiful, picturesque, and sometimes startling scenery.

The city of Portland is situated in a most picturesque location, lying on both sides of the Willâmette River, which is spanned by several fine bridges, and having Mount Hood for a magnificent background. That Mount Hood is a very effective background no one can doubt for a moment, and the fact that this city possesses it is a source of pride to the humblest citizen.

This mountain is located in the Cascade Range, 25 miles south of the Columbia River, and is 12,000 ft. high. It can be seen from most parts of the State, and in fact any part of the State without a view of Mount Hood may be set down as of small importance. It is 60 miles from Portland, but may be annexed in the immediate future and thus brought nearer, for the Portlanders already consider it their peculiar property. That it is lovely is at once to be seen, and it is no wonder they are proud of it ; but just what they have done in its construction no one can say—this must be left to the imagination ; but there is no doubt they did something. They once tried to illuminate it, on one evening in 1885, which was, of course a brilliant idea, but the mount evidently did not approve of the performance, and sent down an avalanche about 4 p.m., setting off the whole thing—fireworks, red-lights, &c.—by daylight. The day selected was our great national holiday, July 4th, and the failure was a great disappointment to the Portlanders. But if Mount Hood thought the matter would rest there, it knew little of its neighbours, nor of the spirit of Portland people, for in 1887 they renewed the effort, choosing the same day in that year. The party started at 5 p.m. on the 2nd of July to ascend the mountain, in full hope of attaining the summit. After terrible hardships they camped in the snow about 7000 ft. up, and the wind blew so that a fire was impossible. Nearly frozen and much exhausted, they started again on July 4th at 5 a.m. with 100 lb. of red fire. A location was fixed at 10,000 ft. elevation and two men left to illuminate. From this point they had a view of 150 miles of their State in one direction, which of itself must have been quite satisfactory. After being in grave doubt as to whether their signal would be seen on account of the clouds and mist, they were rejoiced to behold a red light in Portland, and at once set off their fire in response. The whole city was in a great state of excitement, and so were several neighbouring places. Everything was lighted up in all directions, and the party, after great hardships, including a fall into a crevasse, at last returned in safety. This sort of thing may do for once, and it is not recorded that it was ever repeated.

The mountain also serves the people as a standard of comparison. "Not so high as Mount Hood;" "Not so grand as Mount Hood;" "Easier to climb than Mount Hood," &c. But the acme of these comparisons was reached by the "oldest inhabitant," who was dilating on the length of time he had passed in Portland and the marvellous growth of the country. "Why, gentlemen," said this elegant and eloquent Ananias, "when I first came out here, Mount Hood was nothing but a hole in the ground and now look at it!"

Portland has nearly 100,000 inhabitants, and strikes the visitor instantly as a place of unusual energy with a bright future before it. The hotel where the party stopped is called the Portland, and is built on three sides of a square, leaving a courtyard in the centre, and having the entrance to it on the fourth side. It is a very well kept house, and everything connected with it is first-class. One little precaution was suggestive : each guest has a meal ticket bearing the number of his room, and is liable to be asked for it at any meal. This prevents any dead-beating. Portland is thoroughly supplied with electric railways, and has many fine business streets. The one shown in Fig. 85 was selected rather because a photograph of it was available than as the best street in the city, although it is one of the best and contains many fine buildings—among others a fine book-store, kept by Gill and Co., where the latest literature of the day, both American and foreign, can be had. As a mark of the enterprise of the West it may be noted that almost every city of importance has a fine book-store, and it naturally follows they will have fine schools. Portland's school buildings are among the finest in the United States, and are quite numerous, while its public libraries are buildings any place might be proud of, and are not exceeded in beauty by more pretentious Eastern cities.   Portland has also its Chinese quarter, and in this is a Chinese restaurant.

One other race flourishes in Portland, and seems to thrive there as everywhere, viz., the Jews, and their synagogue, shown in Fig. 86, bears evidence to their numbers and wealth.   The visitor should make a complete tour of the city and note its beautiful buildings, for they are well worth the visit.   He should take a drive to Whitehouse and see Riverview Cemetery, and cross over to Milwaukee, Oregon, returning *viâ* East Portland.

Fig. 85.   Street in Portland.

Another very pleasant ride is to Vancouver, Oregon, to see the garrison. A visit should also be made to Mount Tabor, to the City Park, and to Portland Heights, while another very interesting trip is to the mouth of the Columbia River and Astoria.

Many of these trips can be taken in the electric cars at small expense, and the Astoria trip can be managed nicely by boat.

A visit to the Willámette Falls (accented on the second syllable) shows one source of Portland's wealth.   The picture (Fig. 87) gives a good idea of their size and height ; moreover, the falls are always available, never

FIG. 86. SYNAGOGUE, PORTLAND

FIG. 87. WILLAMETTE FALLS, COLUMBIA RIVER.

running dry in summer, nor freezing in winter.   They are but 13 miles from the city, and have over 50,000 horse-power as a minimum.   The people are preparing to transmit this power by means of electrical appliances and make it available in the city, but already large mills have been built around the falls and are in successful operation.   The records of the Clearing House of Portland for 1891, show 102,570,167.36 dols. for that year, and it was stated that Portland has more than one citizen whose

FIG. 88.   ONEONTA GORGE, COLUMBIA RIVER.

assets are over 16,000,000 dols.   We did not, however, have the pleasure of meeting him or them.

The final trip, which no one should neglect, is to the Dalles of the Columbia.   This is accomplished by rail, starting in the afternoon, and returning by boat the next day.   One requires patience, however, for the boat is quite deliberate, and stops at any point where it is hailed from shore.   The distance is but 88 miles, and the exertion is well repaid by the trip.

The railroad passes in full view of the lovely Gorge of Oneonta, shown in Fig. 88, and directly in front of the Multnomah Falls (Fig. 89), which

FIG. 89. MULTNOMAH FALLS, COLUMBIA RIVER.

FIG. 90. THE PILLARS, COLUMBIA RIVER.

drop in two leaps over 800 ft. The train stops almost in the spray at the foot, and the passengers have an opportunity to descend and go to the bridge shown in the picture. The road also·passes through the Pillars of Hercules (Fig. 90), and through many other places of great beauty and grandeur. On arriving at Dalles City, which is one of the older settlements on this river, a visit is made to the Dalles (Fig. 91)—a distance of but five miles. Here the river runs for $1\frac{1}{2}$ miles through a narrow gorge about 300 ft. to 400 ft. wide, narrowing in one place to 60 ft. The rock is basalt, and the scene is wild and romantic, as though nature had made a tremendous effort to stop the course of this mighty river, which the river resented by breaking through at these points, showing how impossible it

FIG. 91. THE DALLES OF THE COLUMBIA RIVER.

was to restrain the desire for its natural rest in the calm Pacific. The best plan for the tourist is to go from Portland, as stated, in the afternoon, and take a state-room in the boat on arriving at Dalles City, so that he may rise in the morning at such hour as is convenient, since the boat leaves quite early.

The river runs for miles between rocky banks, which in some cases are precipitous, rising to the height of several hundred feet, but at the entrances to Hood River on the north, and White Salmon River on the south, pleasant valleys are seen in the distance. The former river, as its name implies, is the result of the melting snows from the mountain which forms a feature in the landscape at many points in this trip. But Mount Hood is not the only mountain lending beauty to the view. Mount Adams comes in occasionally as an assistant, sometimes with most delightful effect. The shore changes in the descent of the river, which is quite rapid, and the banks are very densely wooded in spots, while in others only bare rock is visible—in fact, the Columbia in many places reminds one of the Hudson River, not alone in size but in scenery. At the upper cascades is a reminiscence of the early settlers, namely, the block house shown in Fig. 92 built some 40 years ago to protect the settlers from the Indians, and garrisoned by Lieutenant Sheridan, afterwards the famous Phil Sheridan. At this point the travellers are carried by a narrow-gauge railroad for six miles to the lower cascades—for the river is not navigable owing to the rapids—and re-embarked for the trip down on another steamer. The Government is engaged in building a canal around these cascades, and each inhabitant has a different opinion of its value, as well as its cost to the

FIG. 92. OLD BLOCK HOUSE, COLUMBIA RIVER.

present time, for it has been under construction for some years. The best average result the writer could obtain from a conservative Oregonian fixed the sum expended at 4,000,000 dols., although one man, who vowed it would be of no use, set this as high as 9,000,000 dols. After passing various scenes of great beauty, the traveller's attention is drawn to a strange object, which is known as a " fish-wheel." These unsightly objects are moved from point to point, and are usually anchored for a time at some cascade. The wheel dips into the stream when in action, perhaps to a depth of one half its spokes. The current affects it as it would an undershot waterwheel, and any salmon once in it is gone; there is absolutely no escape; he is slowly hoisted out of water by this engine of destruction, and on its downward revolution he is shot into a trough and deposited in the bottom of the scow with a deliberation and certainty that seem almost fiendish, the slowness of the movement serving to render the action all the more cruel. Then, disregarding the poor captured fish, the wheel slowly plunges into the water for a new victim, which it rarely fails to find. From time to time the fish are collected, strung on a wire, and fastened to a small buoy and started down the stream with a tag attached, to show what wheel has shipped them. Each cannery has a small steam launch manned by eagle-eyed men, who proceed, on sighting a buoy, to capture it. The fish are taken to the cannery, and the wheel is credited with the catch. It is quite an interesting sight to see these launches darting out in pursuit of the buoys. On reaching the cannery the fish are taken in hand by the Indians, who proceed to clean them, when they are thrown into boiling water, and after passing under the chopper are put into cans which have a vent-hole in the top. These cans are put into a steamer and thoroughly cooked, the temperature being regulated by a thermometer. After this the vent-hole is carefully sealed, and the can is varnished and labelled for the market. The whole process is a very cleanly one, great care being taken to prevent any solder entering the can, and the washings of the fish being frequent. The cannery visited by us was called the Warrendale, and is one of the best on the river. No wonder there is such a demand all over the world for Columbia River salmon.

The captain of our steamer secured some fish that were alive, and had them sent at once to the kitchen. When they came on the table they were the most appetising things imaginable. The reader may judge of this when he is told that one lady of our party, and quite a delicate one, ate three salmon steaks, and others confessed to five ; these only stopped on account of the look of the thing, and claimed that if they had been alone they would have doubled the record. Ordinarily, in New York City, one of the steaks is deemed sufficient for three people.

After leaving the cannery we passed along the river, which abounded in curious formations, such as Cape Horn, shown in Fig. 93. The reader may be able to trace some similarity between the name and the rock, but if he does he will do more than our party could. The captain of the boat proved to be a most genial man, and invited the party into the " Texas," as the wheelhouse is called, and from it they had a magnificent view of Latourelle Falls. The railroad follows the river all the way, and in some places the road-bed is obtained by blasting the side of the cliff. In doing this it was the practice to lower the men in slings so that they could drill the face of the rock.

The people seem to regard the steamboats as their natural servants, and in one case we were amused to see the boat put into a shelving bank where a number of sacks of potatoes had been left, apparently in a reckless manner, and the men proceeded to load them on the deck; for the owner, finding the access rather difficult for his teams, had left them at the most convenient point, and the "roustabouts," as the deck hands were called, had to pack them on board by hoisting the sacks on their shoulders. The captain on this occasion exercised a most marvellous self-control, for he had ladies in the "Texas," but he afterwards told the writer that he would

FIG. 93. CAPE HORN, COLUMBIA RIVER.

willingly have laid down a 5-dol. bill if he could have expressed his feelings. One leaves the Columbia River with great regret, but the Willámette is also beautiful, and the short sail from its point of entrance into the Columbia to Portland was thoroughly enjoyed by all; the day was marked with a red letter in our calendar. The Indians in this section talk what is known as "Chinook," which seems to be something like Volapuk, being a composite of English, French, and any other convenient tongue. If you desired to state that the cashier of the Chinook Banking Company, Limited, had followed his more enlightened types and fled to Canada, you would not

say he had absconded, but he had "kåp-swal-la klåt a wa," the advantage
being that by the time one has pronounced these words, the absconder is
safe in Her Majesty's Province. No wonder the Supreme Court of the
Chinooks is blocked in its endeavours to mete out justice when they must, in
obtaining a jury, tell the sheriff to draw a "men tik-eh kum-tux kón-a-way
me såh che, pe mam-ook klóshe kopa-til a kum." A lawyer there would
take longer to sum up than any shining legal light of the New York Bar if
he had to frequently say, as is common, "Gentlemen of the jury." Is it
anything strange if this people hold the marriage relation lightly when a
husband is known as "Ikt man kwónē sum mit-lite ko-pa-ikt kloóch man,"
and the wife as "Klooch-man mitlite kopa ikt man"? Surely Welsh
must be a kindred tongue. One's imagination pales at the thought of
swearing in Chinook. It must surely be something to loosen the plaster on
an ordinary wall.

With no ordinary subject for reflection, and one which might readily
produce a first-class nightmare if we thought in Chinook, we retired to our
hotel, and the next morning were speeding towards Tacoma and Seattle,
from which latter place we were to embark for the trip to Alaska. But
before entering upon this Alaska journey we were to spend several days at
these cities in the order named, and to enjoy to its fullest extent that
abundant and cordial hospitality which every Eastern man of the right sort
experiences at the hands of his Western brother. The distance is but 145
miles, and the trip proved a most delightful one, our route lying through
some of the most beautiful and fertile valleys imaginable. Thirty miles
from Portland the train is run upon a deck ferryboat and transferred across
the Columbia River, which is very swift at this point, to Kalama, and the
traveller passes from the great State of Oregon to the State of Washington,
a State only lately formed from a Territory, but inhabited by people whose
energy and keenness are unequalled by any encountered in our travels.

By some mysterious means they were fully informed of our coming,
and had out their committees to meet us—business men of standing, jolly,
cordial, and hospitable to any extent. They meant we should have a most
delightful opinion of Tacoma, and it is hardly necessary to say their wish
was fulfilled. On the evening of our arrival we were entertained at the
Commercial Club, and the next morning were taken to see the city. As a
tribute to the enterprise of our country, this city and Seattle are probably
super-eminent. One must be extremely careful to name them together and
in proper order, for the jealousy between these queen cities of Puget Sound

is carried to an almost ridiculous extreme. The writers therefore visited Tacoma and Seattle on the way to Alaska, and Seattle and Tacoma on the return. So at present, being on the way up, we will take Tacoma first. In 1880 there were 1098 inhabitants here, and in 1890, 35,858. At the time of our visit there were more people there—just how many more the writers cannot say, nor would if they knew; for the respective census of these cities is one of the tender points at issue between them. Tacoma, as appears from a glance at the map already given, lies on Puget Sound, at the head of this great body of water. It is located on a high bluff, and was at one time the terminus of the North Pacific Railway. A very good general view of the city is given in Fig. 94. If, however, the reader thinks this to

FIG. 94. THE TOWN OF TACOMA.

be a mushroom town, he is very much mistaken. It is full of life, and has come to stay. The main street, Pacific-avenue, abounds in fine buildings, and many more are under construction. A glance at Fig. 95 will show the character of these structures, which will compare favourably with any in the larger cities. In one year the buildings erected, reached a value of 5,901,000 dols. The electric roads run in every direction, and the traffic on them is enormous. Like all growing towns in the United States, the benefit of popular education is fully recognised, and Tacoma has several fine school buildings. The Opera House is also a beautiful structure, handsomely decorated, and has the accessory of elegant scenery together with a fine stage and ample auditorium. Being on the Sound, and having extensive water facilities, the shipping interests are naturally prominent, and many vessels are collected at the wharves, not only coasters, but steamers sailing to China and Japan. Naturally these various craft require good dry docks, and there is a full supply of this accommodation.

The waters of the Sound are full of fish, and this industry is in a very flourishing condition. The amount taken at one haul brings to mind some of the fish stories of the Bible, and in evidence Fig. 96 is offered to the

Fig. 95. Pacific Avenue, Tacoma.

Fig. 96. Fishing in Puget Sound.

reader. Tacoma can also boast of a church which, though of comparatively modern construction, has the oldest tower in the United States, for it is a tree which, at a moderate computation, started at least 400 years ago.

Early on the morning after our arrival, our hosts, who had reluctantly taken leave of us at a late hour the night previous, mustered in force, and took us to the Northern Pacific shops, located about five miles from the city. and which had just been completed. There had been great strife among the various places for the location of these shops, but Tacoma won. and the result is a series of shops that any railroad might be proud of, containing the latest and most improved forms of machinery. The exterior is shown in Fig. 97. At the time of our visit these shops were running, and turning out fine work. They cost 1.000.000 dols., and pay 40,000 dols. in monthly wages. Among other institutions Tacoma has 19 banks, which seem to be thriving. Returning from this trip, the party were taken to the permanent Exposition Building and shown some of the resources of the State, which were a great surprise, for they appeared to include about all the useful minerals known, and fruit of a character and size that were

FIG. 97. NORTHERN PACIFIC RAILWAY SHOPS, TACOMA.

simply astonishing. We next went to a smelter which had an output of $55,000 dols. in 1891, and paid 75,000 dols. in wages. The people of Tacoma, with wise forethought, were laying out a park in a location overlooking the Sound. It contains 700 acres, and was in that stage of development where they were pulling up stumps and making drives. With the usual energy of the people, one visiting it three months after us might expect to find completed roads, fountains, and statuary. The outlook from this park will be a most delightful one, for the visitor can gaze for miles along the Sound from the point of the bluff on which it is located.

Like Portland, Tacoma has its mountain. and a fine one it is, rising 14,400 ft. into the clear sky, its top crowned with eternal snow. A view of the summit of this mountain, from a photograph taken above the clouds, is

T

given in Fig. 98.   There is, however, one drawback to the pride which the
people of Tacoma feel in the possession of this great natural advantage, and
that is that it is also claimed by their rivals in Seattle, since it is located at
nearly the same distance from each city.   Their only satisfaction is that

FIG. 98.   MOUNT TACOMA.

FIG. 99.   HOP FIELD, WASHINGTON.

they call it Mount Tacoma, while the Seattle people scorn this title, and
call it Mount Ranier.   Woe betide the stranger who forgets the distinction
and fails to give it the proper name in each city !   There have been cases of
mysterious disappearance which are attributed to this blunder ; and so, after
speaking with great pride of Mount Tacoma, we bade farewell to our genial

hosts and started for the neighbouring city, where we spoke with equal delight of Mount Ranier. Our journey was most delightful, and the road ran through a lovely valley filled with growing towns, while on either side of the track we were pleased with the sight of great hop-fields, one of which is to be seen in Fig. 99. A good description of this industry is given in the paper issued at Tacoma, and it is here given in full :

The profits realised from the careful cultivation of hops in the State of Washington are large. The average yield of hops in New York and in the hop districts of England and Germany is not more than 650 lb. to the acre, while the average of Washington hops is 1600 lb. to the acre. I have harvested 168 tons off 170 acres, and in several instances coming within my own personal knowledge, over 4000 lb. per acre have been harvested. The cost of producing hops in New York is not less than 15 cents per pound, and in England an average of at least 18 cents per pound is reached. In the State of Washington the average cost is less than 10 cents per pound, so that Washington hop-farmers frequently sell at a profit, while others, producing at a higher cost, are selling at a loss. For five years, from 1886 to 1890 inclusive, I have kept accurate account of the production of one farm of 61 acres which I have in hops. During this time there were grown and marketed 574,602 lb. of hops, which sold at an average of 17 3.5 cents per pound, aggregating 101,129.95 dols. The cost of production was 9 cents per pound, or 51,714.18 dols. The net profit, therefore, from the 61 acres for five years was 49,415.77 dols., an annual average of 9.883.15 dols., and a yearly net profit per acre of 162.02 dols.

Concerning the crop of 1891 I would best repeat what I said in a bulletin to the trade issued September 24 : "We selected and picked separately the yield off 816 hills of hops growing on a compact lot of ground of the Puyallup Hop Company's yard at Kent, and obtained 214 boxes of hops, which weighed 24 lb. of cured hops per box. The hills of hops are 7 ft. apart, with but two vines to the pole, and one pole to the hill. It takes 889 hills to make an acre. At the rate this compact block of land, consisting of 816 hills, yielded, an acre would yield 5592 lb. of dry hops. Knowing that yield was enormous, and that this statement might be questioned, I have taken special pains to prevent any mistake. One man was detailed to keep the account as the hops were picked, reporting meanwhile to Mr. Arthur Titus, of Kent, who was the field superintendent, and who makes affidavit as to the correctness of the count of boxes. I can myself certify as to the area of the land. No extra care of these hops was taken, nor was there any fertilising material used. This is the greatest yield on record, and entitles Washington to be known as the banner hop district of the world, and the Puyallup Hop Company as having produced the largest yield on record for a single acre, and a further record of having just harvested 126 tons off 125 acres."

About 35 miles of ride brought us to Seattle, which, like Tacoma, is on Puget Sound, and, like it also, has grown in a most marvellous manner. In 1859 the place appeared as shown in Fig. 100, but in 1880 it had 4533 people, and in 1890 the census was 43,914. In 1892 it had more, but on this point the writers will again preserve a marked silence as to numbers, as they do not wish to offend either city, since the hospitality of both was so lavish and so enjoyable. Seattle is also located on a bluff, and· by this it is not to be understood that it means anything more than the natural location.

On our arrival we gladly sought an hotel for a rest previous to our journey
to the far north ; but this was not to be, as we speedily found out, for
Seattle had a reception committee to look after strangers, and they were
just as bright, and just as kind, and just as cordial as those we had recently
left in Tacoma They were proud of their city, and wanted us to see it and
join in this pride ; so we started to look at the buildings and their system of
cable and electric railways.   The city was destroyed by fire in 1889, and the
newer and finer city has risen phœnix-like from the ashes.

FIG. 100.   SEATTLE IN 1859.

The name is taken from an old Indian chief, and had it been possible
for our enterprising committee to have performed the miracle of the witch
of Endor, no doubt we should have seen him ; but, failing in this, they
proceeded to show us his daughter, a woman of great age as well as of large
girth.   Starting then from Hotel Ranier, which stands on the summit of a
high hill and commands a most beautiful view of the Sound, we descended
five terraces before reaching the water front, each terrace being occupied by
a fine broad street filled with imposing buildings.

That the reader may fully understand that these buildings will bear
comparison with those of any city in the world, he is invited to inspect
Figs. 101, 102, and 103.   The Pioneer Building, in Fig. 101, stands on the
site of the little wooden building shown in Fig. 100.   The interiors are quite
in keeping with the elegance of the exterior.   Many are wainscoted in

FIG. 101. PIONEER BUILDING, SEATTLE.

FIG. 102. NEW YORK BLOCK, SEATTLE.

coloured marble, and all have quick elevators. The County Court House stands on a summit not far from the Ranier, and forms an imposing feature in the landscape. A view of it is given in Fig. 103. The year after the fire no less than 13,547,000 dols. were expended in buildings and public improvements; the jobbing business for that year was 31,000,000 dols. and the output of the manufactories 11,000,000 dols.

A short distance (about three miles) from the city is a beautiful lake, some 18 miles long, known as Lake Washington. This is one of the summer resorts of Seattle, and having chartered a steamer as soon as our

FIG. 103.   COURT HOUSE, SEATTLE.

arrival was noted, our kind entertainers took us for a trip on the lovely sheet of water, which lies like an immense mirror at the foot of the surrounding hills. We sailed among the islands, landing at one, and finished the day by a dinner prepared at a lakeside house and served on the balcony.

It was at Lake Washington these citizens endeavoured to have the United States Government establish a navy yard, so that the vessels, although within easy reach of the ocean, should still be floating in fresh water. An outlet was planned into Puget Sound, and will, no doubt, soon be constructed; for the size of an enterprise never daunts our Western inhabitants. A public park is being laid out—indeed, is already laid out to

FIG. 104. LOGGING CAMP SEATTLE

a considerable extent—along the shores in lovely groves, and on Sundays 5,000 to 10,000 people assemble there, being brought by various lines of street railways.   These roads run through woods and among stumps in the trip, for the country is still wild.   It was stated that a bear and a mountain lion were seen on the track the day before our arrival.

The United States Government, however, did not take kindly to the Lake Washington site, and selected one at Point Orchard, some 16 miles from Seattle.   Probably the pressure from this city and Tacoma was such as to lead to a middle course ; at all events, the Government has located there and commenced construction.   It has great possibilities, and is a well-chosen site.   We were taken there by one of our hosts in his beautiful steam launch, built by Herreshoff; this boat can readily make from 16 to 20 miles per hour, and the trip was thoroughly enjoyed.

Naturally timber cutting is one of the great industries of this section of the country, and a logging camp may be appropriately referred to here, as the largest saw mill in the country is opposite the city of Seattle.   Fig. 104 shows several views, especially the manner in which the logs are delivered down the shute, into the water, striking which produces the effect seen in the lower right hand corner of the illustration.

The canal necessary to connect Lake Washington with Puget Sound would not be a difficult matter either financially or from an engineering standpoint, and its construction would put the mills and factories already on the lake shore in easy communication with the ocean.   It would, more-over, stimulate the construction of new industries.   The advantage of fresh water is a great one, especially when it is considered that the *teredo navalis* infests these waters, and will destroy a pile in two to three years.

But there is another and far more important canal, about whose con-struction Seattle and Tacoma will surely agree, and in advocating it they will have the support of all the cities on the Pacific Coast, as well as many on our Atlantic side, and that is one across the Isthmus.   Every American who cares for his country should advocate this on all occasions.   Its value to the nation is inestimable, nor should any political considerations be allowed to hinder its construction.   It is as important to the United States as its navy, and it cannot be made any too soon.   We had a hint of its usefulness in the threatened Chilian trouble ; but, laying aside any warlike aspect, and viewing it for purely commercial purposes, it would put the entire eastern coast of our country into quick communication with the western coast of South America, and secure to us the greater part of the trade with the nations there; while for our domestic trade between the eastern and western coasts of the United States it is even of greater importance. To the visitor on the Pacific side of this great country this view is constantly presenting itself, and if our trip had stopped here the benefit to us as American citizens in obtaining this knowledge alone cannot be under-estimated.

The reader must not suppose that there are no other cities on Puget Sound but Tacoma and Seattle. There are none approaching them in size,

FIG. 105. WHALEBACK STEAMER, EVERETT.

but 30 miles north of Seattle, at the mouth of the Suohorinsh River, is located the city of Everett, in age just one year; but, in energy and latent force, a young giant. Steel whale barges are a new industry, but Everett is building them, and a sample is seen in Fig. 105, although the shops are not yet completed. Everett has also the Puget Sound Wire Nail and Steel Company, which intends to supply China and Japan with articles of this character. A large paper mill has recently been added to the industrial works of this place, and it occupies no less than 28 acres. One of the

U

paper machines is 106 in. wide, and one is 90 in. ; both were made by members of the American Society of Mechanical Engineers. Cottonwood, which is largely employed, is very abundant here, and is peculiarly fitted for making the pulp, since it requires, on account of its colour, less bleaching, and the pulp is therefore in better condition than that requiring the use of large quantities of acid.

The clay in this section makes admirable bricks, and the Everett people have been quick to recognise this ; their brickyard is still in the forest, so rapid is the course of events in this locality. Just now there is only time to make a clearing and begin work. There will be time later to fix up

FIG. 106. HEWITT AVENUE, EVERETT, WASHINGTON.

things. That is their idea in everything. As another evidence of this, a view of Hewitt-avenue is shown in Fig. 106, taken six months after the trees had been burned off, and four months after the first lot was sold. As a summary of this wonderful place it may be said it has now, or had a few months ago (which is quite a different thing), five banks, a hotel of good size, and one under construction to cost 40,000 dols., several churches, and 30,000 dols. voted for school purposes. The population is about 3000, and when our arrival was chronicled a committee waited on us, coming first and last nearly 100 miles to find us (for they went to Tacoma when we were at Seattle), to persuade us to visit Everett and be received there. The good-natured president of the Everett Realty Company proved to be a personal friend of the writers, and at his hands we received many courtesies; under his able control this town is "booming" in a most remarkable manner, which will yet cause some of its neighbours to look to their laurels. But, pleasant as all this was, and opening an experience quite new to all, there lay beyond us the "Land of the Midnight Sun," the last piece of territory added to the domains of the stars and stripes. Alaska was to be visited, and that to the writers' minds was the crowning glory of this delightful journey, so we lay down to rest in the Hotel Ranier full of pleasant thoughts of the trip yet to be taken, only wakening in the early morning to look from our front window down on the bay, and to recognise at the wharf the steamer Queen, which was destined to be our home for the next thirteen days.

The reader must understand that the trip of 2600 miles that lay before us is almost entirely accomplished by a route between the islands and the mainland, the former serving as a barrier to the ocean waves. And if, further, the reader refers to a map, he may think, by journeying to Sitka, he will see a great deal of Alaska, yet he will have only reached the extreme easterly end of these possessions, for this country extends 2000 miles west of this place, and comprises a territory equalling one-sixth of the entire United States, or one-seventh of all Europe, or 580,000 square miles, to put its extent into figures. The general coast line is 4750 miles in length, and taking into account the islands and bays, Alaska has 26,376 miles of shore. The furthermost island of the Aleutian Chain is as far west of San Francisco as Bangor, Maine, is east of it. In fact, San Francisco is now the geographical centre of the United States. The most northerly point of Alaska is as far from the most southerly, as Maine is from Florida, and the width of this territory is as great as from Washington D.C., to California. Mount St. Elias is the highest mountain on the North American Continent, being 19,500 ft. above the sea level. The great river, the Yukon, computed to be not less than 3000 miles long, is navigable for a distance of 2000 miles, and is from one to five miles in width for 1000 miles, while its five mouths and the intervening deltas have a breadth of 70 miles. The glaciers surrounding Mount St. Elias are estimated to be 20,000 square miles in extent.

One of the most popular errors extant about Alaska has reference to its climate. The winters of northern and interior Alaska are undoubtedly very severe; but the coast south of the Aleutian Islands—the whole of Southern Alaska, in fact—being under the influence of the Kurisiwo, or Black Current of Japan, possesses in reality a mild climate. A record of 8 deg. below zero is the lowest that has been registered at Sitka in 50 years, and only four times during that entire period did the mercury descend below the zero point.   Fort Wrangel, although farther south than Sitka, is warmer in summer and colder in winter, because it is farther removed from the great equaliser—the ocean current.   The Queen Charlotte Islands, on the other hand, have a very mild climate.

The native population of Alaska, with the exception of a single tribe— the Tinnehs, found in the interior—estimated by the census reports of 1880 at something over 31,000 altogether, is not of Indian origin.   Whether Mongolian, Aztec, or both, remains to be proven.   Persons who have devoted attention to the subject have found much in the language, customs, and arts of the Haidas, the most remarkable of these tribes, to support the Aztec theory, while there is also much to suggest Japanese or Chinese origin.   According to the census reports, there are five distinct tribes, viz, : the Innuit, or Esquimaux ; the Aleuts, inhabiting the Aleutian Islands ; the Tinnehs ; the Thlinkets, of Southern Alaska ; and the Haidas.   Those mostly seen by the tourists are Thlinkets, but at Wrangel there are likely to be some Haidas.

After this instructive lesson in geography we may now take up the thread of our story and chronicle the trip to Alaska.   We left Seattle at nine o'clock on Monday, June 6, on the attractive steamer Queen.   It was a bright and pleasant day, but before we had sailed an hour we were glad to don all our winter clothes.   We stopped for a short time at Port Townsend, and again at Anacortes, where we went ashore and walked along the main street until the pavement terminated abruptly in the original mud. Anacortes has been a town that was " boomed " to death, and has a hotel worthy of a city of 15,000 people instead of a few hundred.   As the afternoon waned we ascended to the hurricane deck, whence we could see in all directions.   About us were the green islands, and beyond them the grey of the mountains.   The water was like a blue mirror, and across it widened the silver path of the steamer ; white clouds swept over the fair heavens, and in the west, among purple masses, the sun was setting in an amber sky, surrounded by tiny fiery clouds, and tinging those above with orange and with crimson light.

At seven o'clock the steamer reached Victoria, and we were at once made aware of the sovereignty of Her Majesty by the contemptuous rejection of our money and the demand of 20 per cent. for exchanging it. Our object of search here was an ice-cream soda fountain, for one of the ladies had a weakness for this beverage ; this we were unable to find, and, after passing through a number of the main streets of the town, we were obliged to offer her the plebeian peanut as a substitute. There are some blocks of fine buildings in Victoria, and we saw also beautiful dwelling-houses surrounded by well-kept lawns. On our return to the steamer we sat for some time on the hurricane deck enjoying the moonlight, and congratulating ourselves on the fact that our many calculations were correct and that we should have the moon during our trip. Vain rejoicing ! We had utterly forgotten in what direction lay our pathway, and that we were to visit the regions where Apollo reigned supreme, the " Land of the Midnight Sun." We left Victoria at six the next morning, and at breakfast time we were steaming along through a dense and chilly fog, while the sea, winking in the rain as if with Argus eyes, was as level as a floor. The morning was passed pleasantly with reading and games, and when we emerged from our state-rooms at four, after a short nap, we found the sky clear and the air soft and warm. Stretched in our chairs, we basked delightedly in the sunlight, and enjoyed the lovely scenery. On either side green, pine-clad hills sloped to the water's edge, and behind them the snow-capped peaks, while tiny feathery islands dotted the water. At sunset the small gap on the horizon between the lines of cloud-hung cliffs was filled with golden mist, through which we seemed to see a fairer range of hills beyond, and as the sun sank lower the white masses of fog which had settled on the very highest peaks became a soft rose-pink, and all the silver sky blushed to where the moon in the east waded through dark driving clouds.

At one o'clock a point was passed where no sheltering land hid us from the ocean, but even here the swell was so gentle that few were awakened and none were uncomfortable. June 9 found us quietly steering through a channel bounded by mountains—the nearer ones green and wooded and cut with foaming cataracts, while behind were the eternal snows. Now and then came glimpses up sheltered coves and cañons ; while little wild ducks fluttered over the glassy black water, shivering it into crinkling light, and diving in a line of bubbles to rise quietly some rods away. A great white bird flew across the water to the hills, and a quivering line of reflection followed beneath him. Early in the morning a little boat was seen a short way from one of the shores. At first we thought it might be only a log, but a man in the stern waved his hat, and as it drew nearer we could distinguish the forms of six or seven people. A little later a canoe with high ends and a square sail passed, filled with Indians, two of them wearing scarlet blankets, making a brilliant spot of colour. At sunset, or about nine o'clock, we reached a great open bay where three green islands floated in a pink mist beneath a sky of vivid gold, and just where the sun made a pathway of brazen flakes across the waves rose a great purple promontory— Alaska at last!

On June 10 we beheld a line of scraggy wooden buildings, many of them raised on piles; also some log cabins, and a few tents, with here and there groups of Indians; while towering over everything were the wonderful totem poles, and this was Fort Wrangel. We all went ashore at the dock shown in Fig. 107, where the Queen is seen made fast to it, and visited the shops, in which were sold furs, antlers, and the spoons with

FIG. 107.   FORT WRANGEL, ALASKA.

which the Indians eat; these spoons were really great scoops of horn or wood, carved with curious wide-open eyes and other strange symbols.

The most interesting thing in Fort Wrangel is, of course, the totem poles, and these we next visited, going from one to another, and obtaining some very good photographs of several curious ones. These totem poles, varying from 20 ft. to 60 ft. in height, and 2 ft. to 5 ft. in diameter, are said to have been objects of worship, and also to have served as the family crest, for there is great aristocracy even among savages. When an Indian marries he takes the crest of his wife's clan—for they never marry in their own—and it is placed on the totem pole. These Indians would receive high praise from the progressive women of the present age, for it is the rule for the wife to have charge of everything of a business nature, and no man makes a transaction unless she agrees. The clans are called the Eagle, the Raven, the Whale, the Wolf, &c.

One of the oddest-looking poles has a great bear "treed" at the top, and in two places on the sides huge footprints are cut all the way up

FIG. 108. TOTEM POLES, FORT WRANGEL.

FIG. 109. INDIAN GRAVE, FORT WRANGEL.

(Fig. 108). Another represents a demon devouring a fish who had paused *en route* with his dinner half swallowed. Others are combinations of fish, flesh, fowl, and imagination. All are weather-stained and moss-grown, and

have the appearance of great age. One Indian grave is marked by an
enormous dog (Fig. 109). Longfellow thus speaks of the totems :

> And they painted on their grave-posts
> Of the graves yet unforgotten
> Each his own ancestral totem,
> Each the symbol of his household—
> Figure of the bear and reindeer,
> Of the turtle, crane, and beaver.

At six o'clock the stopping of the engines caused excitement, which
increased when it was known that we were approaching a steamer and
were going to be brought close up to her. Gradually the steamer, which
proved to be the English ship " Islander," drew nearer and nearer, gay with
fluttering handkerchiefs, until at length the two vessels were side by side,
and a fender was lowered to prevent them from rubbing against each other.
After about ten minutes, when newspapers had been given and hands
shaken with the assistance of an umbrella, each shaker grasping one end of
it, the steamers separated, both whistling, while everyone cheered as the
English flag dipped three times in answer to the stars and stripes ; and in
a few minutes the Islander was a white spot in the distance.

And now we began to realise how far north we had come, for at
midnight one could easily read on deck, and it was hard to sleep in the
grey twilight. Early in the morning we were wakened by steps above us,
and excited cries that we were at the North Pole.

Looking out of the window, and, in spite of the intense cold, remaining
there some time, we were transfixed by the wonder of the scene. On either
side great mountains covered with snow and hung with wreaths of white
fog plunged sheer to the water's edge. The bay was slate-colour, and on it
floated great jagged masses of turquoise-blue ice, which cast a bright
reflection on the glassy stream.

At seven o'clock we all went out on the hurricane deck and beheld
between the clustering mountains the blue spires of the Taku Glacier rising
100 ft. into the air. On account of its exquisite vivid colour, the Taku
Glacier was to us the most beautiful of any we saw. The bay in front of it
was filled with masses of ice, which clinked coldly as they struck together ;
then came the sound of rushing water, which poured down the rocks from
the snow banks, and filled the air, and now and then came a boom like
distant cannon from the glacier, a cloud of powdered ice rose, and a piece
detached itself and sank into the water. A large black bird with ragged

wings circled over the bay, and finally lit on a peak of ice. His hoarse cries added the last touch of desolation to the scene. Although we had often read descriptions of Alaska, and had thought that our eyes were prepared to see this, yet it seemed to strike our minds with great surprise, and we almost felt that it was a dream, and could hardly realise that this wild scene was a reality. To the left a sluggish mud glacier poured its yellow slime into the water, and beside the steamer a little boatful of men were procuring the ice which was to last the Queen for her next two trips. One man standing in the bow caught the piece he desired with a hook, and when his companions had drawn it to them a net was lowered, which they put under the ice, and it was then swung into the hold of the steamer.

The next point of interest was the Tredwell Mine, situated on Douglas Island, near Juneau. Strange as it may seem, the mill at this mine has the largest number of stamps in any mill in the world, no less than 240 being employed. The mine is worked in the open cut, and the ore is of low grade, running from 6 dols. to 18 dols. per ton. It was discovered in 1860, but has been worked in earnest only since 1885. The ore contains pyrites, which were mistaken by many of the party for free gold. One elderly person carried at great trouble—for she was quite stout—a large piece on board the steamer, thinking she had a " find" rich enough to pay all the expenses of the Alaskan journey, and she looked so happy over it all that no one had the heart to undeceive her.

In the afternoon we reached Juneau, and having been informed in a little pamphlet, copies of which were distributed at Fort Wrangel, that this was the metropolis of Alaska, containing fine jewellery stores, many inhabitants, &c., we fondly anticipated wide streets and fine buildings, and could hardly believe our eyesight when we beheld (see Fig. 110) the straggling lines of tumble-down wooden buildings, and the ragged narrow sidewalks beside a slough of mud. The day was bleak and chilly, and we wandered along the dirty streets pelted by a fine slanting rain. One curious little house hung with skins and bright basketwork was built on piles some distance from the shore, and was reached by a narrow board walk swinging high in the air. As we attempted to cross this uncertain bridge we were considerably startled by discovering, when half way over, that three bear cubs were chained upon it. They proved perfectly amiable to us, but later one of the party tried to pat the head of a cub, and left the entire back of his glove in the brute's mouth as a souvenir. When they were driven aside with an umbrella, we passed between them to the shop.

x

It contained innumerable Indian curiosities—bones, arrow-heads, and carved stones used for witch charms.   A Sister of Charity, next to whom we had the good fortune to be seated at the table, told us a great deal about Juneau and the good work of the Roman Catholic Mission there ; she said that the practice of exorcism for suspected witchcraft was still carried on secretly among the Indians, and that when the victim died he was quietly disposed of

FIG. 110.   JUNEAU, THE CAPITAL OF ALASKA.

without the knowledge of the Government officials.   We were also told the sad and horrible story of an Indian uprising some time ago.   The two tribes at Chilcat had trouble, and when they finally came to blows one man was killed.   The officials then succeeded in interfering, but both tribes alike demanded the sacrifice of a man to appease the ghost of the dead one.   Of course the Government authorities refused to allow this, and they finally sent to Sitka for assistance and a higher officer.   He came at once, but so fierce was the feeling among the Indians that some terrible evil would fall on them unless they were allowed to sacrifice the man, that nothing would stop them, and they became uncontrollable.   With her brown eyes running with tears the Sister told of the horrors of that day, when the poor wretch was chosen by lot, and, having been taken without the city, was there tortured to death. " He did not die until sunset," she said, " and all day long we heard his screams of agony."

The population of Juneau seemed to consist almost entirely of Indians, traders, and a few miners and hunters. The squaws squatted along on the sidewalk and in the open doors, holding their babies, and surrounded by dirty children and dogs. We were very much startled by one woman, who shoved her head from her blanket and, peering closely at us, disclosed a brilliant crimson face. A little later we saw another whose face was perfectly black, and on inquiry discovered the reason of this strange custom. It is hardly to be believed, but the truth is that they do it—first, because they think it becoming ; and, second, to preserve their complexions from sunburn. The weird effect can hardly be imagined ; the woman, whose long black hair hung over a scarlet face from which gleamed snow-white eyeballs and teeth, reminded one of a flayed horror from the " Inferno."

In spite of their hideous paint and their wild costumes, these Indian women excited our deepest interest and pity on account of the intelligence in many of their faces ; and their spiritual wants have not been forgotten, for the Presbyterians have a struggling mission here.

Another incident enlivened our visit to Juneau. In the afternoon a couple of Indians came to the steamer and tried to buy whisky from one of our firemen. The selling of whisky to an Indian is an offence which the Government punishes severely, as it renders the red men like wild beasts, and for some time the fireman refused. But when he was offered 5 dols. for half a glass he yielded, and as soon as the sale had been made one of the Indians displayed an official badge ; he proved to be a policeman, and took him into custody. Of course the fireman was in the wrong, but we could not help pitying him when we learned that he must remain at Juneau until the court met in November, before he could even be tried.

In the evening many boats loaded with Indians came from both shores and circled about the steamer. The canoes were very picturesque, and looked not unlike Venetian gondolas, being very high at both ends. They are hollowed from a single great log, and carved to suit the owner's fancy. Sometimes they contained as many as fourteen men, women, and children, all dressed in bright-coloured blankets, making a very pretty picture as they paddled about on the quiet water. They seemed very jolly and good-natured, and one big boat-load pulled up close beside the steamer and tried to sell a little stuffed bear. The passengers threw down fruit and cakes, to the great delight of the Indians, and when one bag of peanuts hit on the edge of the canoe and broke, a bright-faced savage displayed his knowledge of English by shouting smilingly :

" All busted !"

When, again, an ill-aimed bag of cake fell into the dirty salt water, they would grab excitedly for the floating cakes, which the fat babies seized and devoured at once. We were told that they live on almost anything they can find, even the most horrible refuse, and stale fish is to them a luxury.

There are several gold mines near Juneau, about three miles back in the country, and the writers intended to visit them, but the rain and mud proved to be too much of a hindrance, and the only horse in town was busy hauling stores of provisions, and was not available. From all that could be learned, it appeared that these mines were doing quite well, and their owners did not care to part with them. In this respect they were an exception to everything we saw in Alaska. It was all for sale, and the prices were perfectly ridiculous, sometimes, and not infrequently, eight or ten times their value. It is better to buy furs at Minneapolis, or even in New York City, as you can purchase the Alaskan furs at those places for about one-fourth of the prices asked in Juneau and Sitka.

Sailing from Juneau, we reached Chilcat, with its glaciers, at some unholy matin hour. Here we saw the Davidson Glacier, three miles in breadth, spreading out like a fan. This was the highest point we reached in the trip, being 59 deg. 10 min. 36 sec., and the sun shone 21 hours out of the 24 at the time of our visit.

At 9 A.M. there was a silver streak on the fog-hung water, which spread and brightened until the sun burst forth and drove away the mist, and we saw about us great ranges of mountains, looking like blue ocean waves with their summits breaking into foam. All the afternoon we passed a succession of beautiful scenes. Sometimes the pathway widened out like a great bay, and we seemed shut in by hills, water, and sky, all of the most vivid blue, while the waves sparkled in the sunlight as though strewn with thousands of diamonds. And now the water shrank into a channel as small as a river, where, with the engine at reduced speed, the steamer threaded its way among tiny green islands, with opening vistas beyond, down other streams of green and snow-clad peaks. We passed many Indians, and went so close to the shore in places that we could look far into the woods and mark the

beautiful shrubs and trees which drooped over the banks. At six o'clock we
saw, between a range of green mountains, the white summit of Mount
Edgecombe, its snowy foothills hid in great billows of silver clouds, and
beyond them the azure expanse of the Pacific Ocean. The water was absolutely quiet, and the steamer rode on past beautiful tufted islands to where
the city of Sitka (Fig. 111) clustered by the Indian River in the hollows of
the hills, with the pale green spires of the Russian Church rising in its
midst. In the wide circle of the beautiful harbour, three men-of-war lay at
anchor, and on the broad green above the wharf was a gay and motley
assemblage of sailors, marines, and Indians. As soon as we landed we

FIG. 111. THE TOWN OF SITKA.

walked through the town, stopping to examine the bright baskets and silver
rings and bracelets of the squaws who lined the road. A white-washed wall
made a fine background, and our photographers were extremely anxious for
pictures. Although some of the women "put on airs," giggled in strangling
gutturals, and coquettishly veiled their faces with their shawls, or hid them
in their hands, they were not really afraid, like the Indians of the prairies,
and we succeeded in obtaining two excellent pictures. The little museum
out at the Mission some two miles distant was then visited, and there we
saw curious nets made of whalebone; canoes made so that the oarsman
buttoned himself into the centre just as he would button on a jacket; suits of
salmon skin, and Indian, Russian, and Esquimaux relics of all kinds. One
of the young men from the Indian Mission acted as our guide, and politely
explained everything. The party then accompanied him to the Mission,

where we enjoyed a most interesting performance, which reflected the greatest credit on both teachers and pupils.   First came a drill with American flags, in which the whole school took part, and then some pupils, first a boy and then a girl, stepped to the front of the platform and made a little speech, personating one of the United States and telling of its characteristics.

There are now 50 girls and 100 boys in this institution, of various ages, the youngest being but three and the oldest twenty-two years.   They are taught English branches of education, and also many useful trades, which they practise among their own people.   When the exercises were concluded, we passed out into the front yard, where a band, composed entirely of Indians, played surprisingly well, keeping the most perfect time, and giving us in several instances tunes of the day, showing they are quite up in popular music.   The work done among the Indians is showing most excellent results, and the number of well-dressed and intelligent-looking Indians about the town affords an evidence of it.

The people here claim that these Indians are superior to any others in the country, and are much more easily brought to adopt the ways of civilisation.   It is obvious at a glance that the face of the Alaskan Indian is in many ways radically different from the red men of the south, and it has even been suggested that these people were originally Japanese who were wrecked on the Alaskan coast, and the flat noses and peculiar shape of their faces and eyes seem to give credence to this theory.

We found Sitka a unique and an interesting place, for it has inherited from its Russian ancestry a distinctly foreign air, and in it is a mixture of nations, Indians, Russians, Esquimaux, and Americans, and a babel of tongues, for some of the Indians still talk Russian in addition to their own dialects.   We felt, too, that we were on historic ground, for this city was the theatre of the bloody struggle between the Russian settlers and the Indians, succeeded later by the peaceful arbitration of Russia and the United States.   High above the harbour is the barn-like structure of Baranoff Castle, formerly the residence of the Russian governor, Baranoff, who ruled here with tyrannical severity in the first part of this century.   It is a pleasure to all Americans to think of that day when, through the far-seeing policy of the Hon. Willian Steward, the sale of Alaska had been effected, and amid the cheers of soldiers and people and the booming of cannon the Russian flag fell, to give place to the stars and stripes.   The Government Building is seen in Fig. 111, and the Revenue Office adjoining, while at the end of the main street appears the Russian Church. The buildings between are all trading stores.

On our way back to the steamer we were attracted by a crowd on the wharf, and on inquiring found that the tourists were shaking hands with the Princess Thom, who is the Indian heiress of Sitka, and is said to be worth as much as 100,000 dols. Do not fancy that we saw a graceful brown maiden decorated with beads and feathers ; we beheld a stout woman weighing about 250 lb. and some 60 years old, dressed in an ill-fitting brown waist and skirt, who, however, smiled at the crowd in a most friendly manner.

That night we saw a genuine Indian dance, or rather a series of them. A place was cleared in a large house, and the spectators were seated on benches ; directly in front of them, squatting on the floor, were some eight or ten very cunning little Indian girls, all under six years old. They kept time by pounding on the floor to the beat of the Indian drum and singing a slow chant. The dancers went out of the room, and presently came in one by one dressed so as to imitate some animal, such as a bear, a deer, a porcupine, or a wolf. They also imitated the actions of these animals. The acting was extremely good, and the performance seemed to be a sort of allegory. As nearly as we could decide, they represented two tribes at war. All of a sudden they rushed at one unfortunate chap and proceeded to take vengeance on him. After he had been mauled sufficiently to the tune of an Indian quickstep, amid howls and shrieks, he was induced to open his hand and display an image. This was the signal for a new onslaught, and the poor fellow looked as though he had been trying conclusions with a patent reaper. After this had been repeated three times, he opened his hand, and it was empty, and then they had a sort of love dance all hands around. We understood the doll was an evil spirit, which they had exorcised, and that now both tribes would be at peace for ever. They also had a medicine dance, and restored a dead boy to life. This was very well done, especially by the boy, who took good care not to come to life till the proper time, and then only to do so by degrees. As he came to life he emerged slowly from a fur robe. The music had a sweet, solemn sound, and the whole performance was excellent. The Indians believe in one figure on their doors to indicate the number of the house, so the Government has allowed them to add ciphers ; thus the first house in this somewhat crooked street is 100, and the next is 200, &c.

The high latitude caused many amusing mistakes as to time, and one lady who had been to call on a missionary on shore was quite shocked to find she had called on the poor man at 10.30 P.M., and had remained till 12 midnight. Nothing but the sight of her own watch on the steamer would make her believe this, for she was a lineal descendant of St. Thomas. The houses had curious signs on them ; one read " Elisha Cadwalder, the father of a large family of Christian children." When we finally retired it was in broad daylight, and the state-rooms had to be closed to keep out the sun, although it was about 1 A.M.

The next morning we visited the Russian Church, where a courteous Russian, seeing our interest, showed us the massive crowns with their great jewels which are still worn by the brides and grooms. Also we saw the bishop's mitre of pearls, which contains one great emerald. We then examined the pictures which were sent here 70 years ago by Russian noblemen when the church was built. Most of them are very curious-looking, as the figures are of gold or silver, with hands, feet, and faces of painted ivory. Apart from the others, in a separate room, is an exquisite Madonna, which our guide informed us is fabled to have been painted by Raphael, and he added they had refused 20,000 dols. for it. Whether this be true or false, the faces in this picture are very lovely, with a delicate softness of finish and great sweetness of expression, and if Raphael did not paint it, there is no doubt he would not have been ashamed of it.

We also entered a little Greek Mission Church, where a service was being held ; many Indians were present, and their resonant voices, as they sang the service, were very weird and pathetic. We then walked to the Indian River, and rested for some time in the pine woods on its shores. Here the moss hung from the gnarled trees, and although it was now the middle of June, the branches were tipped with the new green of spring, and the ground was strewn in places with delicate white flowers. On our return to the steamer the whistle was blown to collect the passengers, and as its echoing and piercing shriek resounded, a most hideous wailing, howling, and barking filled the air, ceasing as soon as the whistle paused, and being resumed when it recommenced. We soon discovered that this noise proceeded from the numberless dogs of the Indian village, who thus expressed their decided disapproval of the sound.

In the evening we reached the entrance to Glacier Bay. The hills about us were dark green with hardy foliage, but beyond the gateway the mountains were covered with snow, and behind these the sun was just setting, wrapping them in glory, tinging pink the little ripples of cloud,

FIG. 114. THE MUIR GLACIER.

FIG. 115. THE MUIR GLACIER.

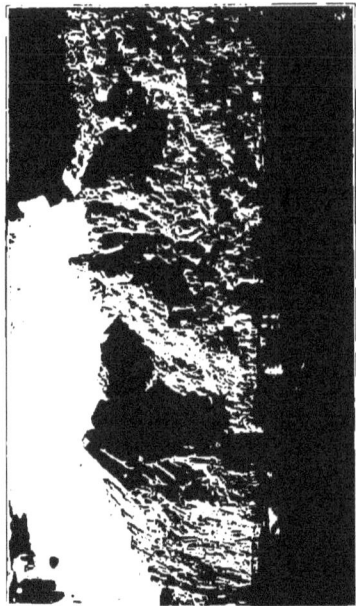
FIG. 112. GLACIER BAY.

FIG. 113. THE MUIR GLACIER.

Y

and its light shading into clear yellow on the horizon, against which the white cliffs were cut sharply, while far in the distance the pale light shone on the level ice of a broad glacier. At midnight the sun was a gleaming spot in the hollow between two high white peaks, and from it one ray of light shot across the waves like a golden arrow. By this time we were surrounded by icebergs, snow-white, blue, and emerald-green, and the water was a black mirror in which they measured their length as they drifted past. Our progress was now difficult, if not actually dangerous, for we were the first steamer of the season to penetrate this frozen bay, and it was here that the Islander had been blocked by the ice and forced to return. Slowly we threaded our way among the floating bergs, with every now and then a crash and jar, and the crunching sound of the ice grinding beneath the steamer (Fig. 112).

At one o'clock in the morning we saw before us the high peaks and masses which the Indians called the "Ice Mountains." This white enormous barrier looming before us through the grey light was the great Muir Glacier (Figs. 113, 114, and 115). The height of the ice wall above the water is about 300 ft., but soundings show that 720 ft. are below the surface; while a third portion is buried beneath moraine material. "Therefore," says Professor Muir, "were the water and rocky detritus cleared away, a sheer wall of blue ice would be presented a mile and a half wide and more than 1000 ft. high. This one glacier is made up of about 200 tributary glaciers, which drain an area of about 1000 square miles, and contains more ice than all the 11,000 glaciers of the Alps combined. The distance from the front, back to the farthest tributary is about 50 miles, and the width of the trunk below the confluence of the main tributaries is 20 miles or more."

The next morning all was bustle and excitement; boats were lowered and parties started on exploring expeditions. This party will be the last crowd of tourists who will ever wander at will on the Muir Glacier, for we brought with us 12 men who were to remain in a little hut on the beach, and whose work it was to put up a number of signs showing the right direction to take and the dangerous places, besides erecting a wooden bridge to cross between two lakes on a strip of ice, concerning which we shall have more to say later. Of course it will be much safer with all these precautions, but we were really glad to be without them, for the spice of danger lent a keen enjoyment to our wanderings, and, somehow, even the greatest scenes are dwarfed by signboards and by the beaten tracks of men.

The ice upon which multitudes have tramped their way appears much less wonderful and even smaller than the trackless waste over which you seem to be the first explorer.

We were among the very last to leave, and as we waited impatiently on the upper deck we could see the earlier parties crawling over the white face of the glacier, looking like little black ants. One athletic young man, whom among ourselves we called "Tartarin," in memory of the immortal Alpine climber, made great preparations for a grand exploring expedition. We saw him start early in the morning with great ropes wound about him, and carrying a steel-pointed staff and other implements in his hand. He took his dinner also, and climbed all day, returning with tales of wonder and beauty, but as the glacier seemed about the same in one part as in another, we consoled ourselves with the thought that we had really seen as much as he had told us he saw. For travellers have been known to magnify a little.

At last it was our turn to go, and we descended the ladder into the little boat, and were rowed across the rocking waves. Near the shore were four men, who held the boat close to a floating bridge. They stood almost all day up to their waists in the icy water, and though they had on high rubber boots, it must have been very cold and uncomfortable. And now came the hard and tedious part of the journey, for we had to climb for two miles over the rocky moraine. This is the hardest walking imaginable, as it is mostly uphill, and the masses of broken rock are embedded in ice. Nevertheless we could not advise anyone to forego the Muir Glacier on account of this hard pull.

Now, as we steadily proceeded, the rocks were more scattered and the ice plainly visible, while beyond us loomed the great white waste which was our goal. We came to cracks several feet deep, and then to a great black hole, into which we threw a stone, pausing to hear it clink, clink all the way down, and "chung!" into water far beneath us. Soon great seams and chasms opposed our advance, and we were obliged to cross the smaller ones and to go around the larger. In a few minutes even the sprinkling of stones was gone, and we trod upon the solid snow of the glacier. Then our further progress became a scene of wonder and delight. It was a perfect day for such a trip, for there was no wind, and the sun was behind clouds, although the sky was vivid blue where it could be seen. The white sea of the glacier streamed away before us in frozen waves to a range of snow-covered mountains far in the distance, and close at our side a little river of the most gorgeous blue had forced a channel for itself and cut across the ice. Great holes as blue as turquoise were all about us, and as we advanced we came to hills of ice and precipitous crevasses. We now formed in single file, and stepped carefully in the footprints of our leaders as the path wound about the slippery sides of chasms and over the crests of snowy billows. At length we saw that all those before us had stopped and were gathered in an excited crowd. We hurried to reach them, and at once perceived the cause of the delay. We had reached a point where but one possible path lay before us, and this was across a little bridge of ice about 6 in. wide and 7 ft. long, which shelved abruptly on both sides to glorious blue lakes, where the ice shone like jewels through unfathomed depths of clear water. A few daring ones had crossed, but the rest were debating as to making the attempt. One lady came boldly to the brink six times, when her courage failed her. She said she had "changed her mind," and then she meekly returned, being satisfied with having exercised that inalienable right of her sex. Formed in line, we firmly grasped one another's hands, and, stepping sideways, crossed the abyss. It is here that the wooden bridge, before mentioned, is to be placed, and it is greatly needed, as a misstep might be fatal. We now proceeded some distance further, and rested awhile on the crest of a wave of ice, surveying the strange and pallid scene ; our eyes travelling over the towers of snow which gleamed in the grey light to where the white hills 40 miles away were cut like cameos against the bright blue sky. Far beneath us we could hear the gurgling of a sub-glacial river, and looking down through a gaping fissure, we saw the hidden torrent many feet below plunging down in a great waterfall.

Awed by the wonder of the scene, we ceased to speak, and over the frozen waste there seemed to rest an almost oppressive stillness, which was at length broken by strains of music. A gentleman of another party was mounted on a pinnacle so far away that he appeared a mere black dot to us, but, on account of the extreme clearness of the air, his sweet voice, and even the words of " America," came to us over the fields of ice.

Our return was, of course, very tiring, but we did not regret the trip, and just as we were leaving the glacier the sun broke from behind the clouds, and the ice glittered in the sunlight. It was very brilliant and beautiful, but we were thankful that this sparkling sea had been in shadow while we were traversing it, on account of the glare, which was dazzling. In the afternoon the captain took a small party of ladies ashore in his little steam launch. As they passed the glacier, a big piece of ice detached itself, plunged into the water with a great splash, and rose to the surface not far from the little craft, causing it to dance wildly on the leaping waves, whereupon the ladies screamed all together, and implored the men to "let them get off! let them get off!" not stating where they desired to be landed. Some of the party returned to the shore and spent the rest of the day wandering about, talking to the few Indians who live in this desolate spot, and in gathering gorgeous red blossoms which grew upon the hills on either side of the glacier.

Before leaving, the steamer approached within half a mile of the glacier, and we saw it rise above us like enormous pillars and spires of marble veined with blue. It appeared in the twilight a great white wall, 400 ft. high, and from three to four miles wide (Fig. 113). The cold was intense, and when our whistle blew, the echo was almost as loud and clear as the original sound.

Then we steered carefully back among the icebergs, and, turning, took our last look at the Muir Glacier. Between the dark ledges of rock its white waves streamed back to the hills against a pale gold sky, and a shimmering glory shone across the sea of ice.

Our return voyage was a pleasant but uneventful one, our only excitement being a little impromptu entertainment by the young people, just before we reached Queen Charlotte's Sound, which is the roughest portion of the trip. Some very clever songs were composed and sung, an Indian pow-wow was given, and the whole affair was exceedingly pleasant.

We had hardly reached our state-rooms when the boat began to plunge in the waves, and one of our poets asserted that he heard a minister chanting sadly,

Rock-a-bye, preacher, in the berth top;
When the wind blows you'll empty your crop.

There really was but a gentle swell, and that only lasted about two hours.

We stopped at Nanaimo one afternoon, to take in coal, and a line of boats tied together and headed by the steam launch, was towed across the water, to visit the town. Some of the passengers, also, went ashore at the coal-yard, and wandering about in the woods, returned with their arms full of beautiful ferns and flowers. The next day we reached Seattle, and on leaving the steamer all expressed their thorough enjoyment of the whole trip, indulging in the hope that they might repeat it at some future time. We cannot think of any journey that would be so restful as this quiet sail. One is entirely cut off from the busy, tiresome world, having just enough change to interest, yet not enough to fatigue either mind or body, and is thus enabled to enjoy a recuperating calm, while the fresh ocean breezes bring a furious appetite, and the meals can be thoroughly enjoyed all the time.

After a short stay at Seattle and Tacoma, we started on the next stage of our journey, and the next great point of interest, viz., the Yellowstone Park. The rains which had been deluging the country made frequent delays necessary, but, on the other hand, they added greatly to the beauty of the scenery, for the forests were green and luxuriant in their foliage, while even the smallest brooks were filled, and rushed by us like small torrents, babbling and purling their welcome.

The Green River follows the railway track for many miles, and the moonlight on its raging waters, causing them to look like molten silver, presented an effect which will long be remembered. The great engineering feat in this road was the construction of Stampede Tunnel, at an elevation of 3000 ft. This tunnel is nearly 10,000 ft. long, and is lighted by electricity. The cañon is marvellously beautiful, and we reluctantly left its contemplation to seek our couches. The next morning we stopped at Spokane for some hours, and the entire party was soon distributed through the city, which, like many other places, proved to all a most delightful surprise. Broad streets and fine buildings were to be seen on all sides. The electric road is everywhere, and no wonder, for by glancing at Fig. 116 the reader will see one of the finest water-powers in the State. This was ascertained to be 30,000 horse-power; thus it enables the street railways to dispense largely with animal labour. And it not only furnishes this power, but also all necessary to light the town. Moreover, several fine mills are in operation on its banks. Spokane has a promising future, and will be heard of as one of the foremost manufacturing towns of the State.

The city is situated midway between Helena, Montana, and the Puget Sound. It is 400 miles each way from a city of any considerable size. The city is located at the Falls of the Spokane River, on each side of the river, the ground rising gradually to a line of bluffs three-quarters of a mile back from the river. On the north side, a broad, level plain stretches for about two miles and several miles east and west from the falls. The business section of the city is south of the stream, the mills and manu-facturing enterprises along its bank.

Spokane is on three trans-continental lines—the Northern Pacific, the Union Pacific, and the Great Northern. In addition, they have the

Spokane and Palouse, the Central Washington, the Spokane and Idaho, the Spokane and Northern, the Cœur d'Alene branch of the Union Pacific and the Seattle, the Lake Shore and Eastern, in direct communication with all points of the surrounding country.

In 1889 a fire destroyed 38 blocks of the business portion ot the city, resulting in a loss of upwards of 10,000,000 dols. Since then the burned portion has been rebuilt, at a cost of between 6,000,000 dols. and 8,000,000 dols. Among the more prominent blocks erected since the fire are the Auditorium, containing a very handsome opera house, with a seating capacity of 3500 people, costing 300,000 dols. ; the Granite Block, Hyde Block, Blalock Block, Rookery Building, Spokane National Bank Building, First National Bank Building. Any of these buildings would do credit to the streets of the larger eastern cities. During 1891, the work of re-building was continued, there being about 350,000 dols. expended in business structures, consisting of 31 new brick business houses.

Within a radius of 150 miles of Spokane, there were produced, in 1890, about 7,000,000 bushels of wheat, and in 1891 the crop is estimated at 17,000,000 bushels of wheat, 6,500,000 bushels of oats, and 4,000,000 bushels of barley. The Palouse country south of the city has an estimated area of 13,000,000 acres of arable land suitable for wheat-raising. The Big Bend country, lying west of the city, has an area of 4,000,000 acres, also suitable for wheat. It is estimated that one-sixth of this country is now under cultivation. With the water power at Spokane, this vast crop will be ground more cheaply there and shipped east or west in the shape of finished product, rather than in the form of grain.

The Edison Electric Illuminating Company have erected a very large power station at the corner of Front and Monroe streets, the building being 65 ft. by 120 ft. The turbines are erected 20 ft. above extreme low water, so there can be no danger of flood interfering with the machinery. These wheels are run under a pressure due to 70 ft. fall of water, those driving the arc machines making 675 revolutions per minute, and those for the incandescent machines making above 1000 revolutions per minute. The company supplies electric light to all parts of the city, runs the elevators in the different buildings, supplies the power to the street cars, printing presses, and all classes of light machinery, &c., all operated direct from the main plant. A long distance telephone system is in operation with nearly all of the mining camps and points of importance in East Washington. The city is rapidly becoming a large flour mill centre, there being four mills in operation, and a new 1200 barrel per day mill soon to be erected. They have five breweries, six lumber mills, three sash, door, and furniture factories, iron works, oatmeal mill, and various other enterprises, which make use of the valuable water power. The value of the water power is estimated at 1,500,000 dols. annually to the city. Spokane has school property valued at about 500,000 dols., with a very complete system of public and high schools. There are seven school buildings. There is also located here the Spokane College and Spokane University, Gonzaga College, and a seminary for young ladies.

The banks in Spokane, with capital and surplus, are as follows :

	Capital. dols.	Surplus. dols.
Washington National	100,000	22,000
Exchange National	100,000	60,000
Traders' National	200,000	104,000
Citizens' National	150,000	18,800
First National	100,000	126,700
Old National ...	300,000	
Browne National ...	100,000	18,000
Bank of Spokane Falls	150,000	125.000
Total	1,200,000	474,500

In addition, there are the Washington Savings Bank, with a capital of 150,000 dols., and surplus of 18,000 dols., and the Spokane Savings Bank, with a capital of 60,000 dols., and surplus of 20,000 dols. The street railway systems consist of 25 miles of cable and electric roads, and are fast increasing. The present water works plant of the city cost approximately 300,000 dols., and, including the franchise, it is estimated at 1,000,000 dols. The present pumping capacity of the plant is approximately 8,000,000 gallons per day.

We now passed from Washington, where we had enjoyed so much hospitality, into the State most lately added to the galaxy of national stars, to wit, the State of Idaho. A stop of an hour was made at Cour d'Alene, where recently so much blood has been shed in an effort to convince the miners that even in Idaho the law is as supreme as in the most civilised parts of the Union. That the lesson has been thoroughly learned there is but little doubt. The men were on a strike at the time of our visit, and came down to the station in large numbers to gaze at us. The railroad follows the shores of a beautiful lake, called Pend d'Oreille.

The last station in Idaho is called Hope, though whether it was the Hope deferred or realised, one cannot state, but as the time changes here and we find we have lost an hour, it certainly was not deferred, but advanced. Here there is a farewell view of the beautiful lake, at the upper extremity of which the road ascends Clarke's Forks, a very rapid stream, and after roaring through various gorges, reaches the Bitter Root Mountains ; in one place, environed by a succession of high mountains, is a lovely piece of landscape known as Deer Park. (See Fig. 117.)

The country now began to change in character, and although still mountainous, yet there were stretches of plains to be seen on both sides, and many herds feeding. We had passed again into a new State of such size that it contains within its borders 1,000,000 more acres than the entire States of New England combined, and as many square miles as New York, Pennsylvania, and Illinois together. This was the great State of Montana, containing 40,000,000 acres of land, or 143,776 square miles, with a population of 131,769.

The first view of moment is at Horse Plains (shown in Fig. 118). We passed over several trestles of nearly 1000 ft. in length, and sometimes over 100 ft. high, and through sections of country where fierce struggles had occurred with the Indians in the early days, until we reached Mullan Tunnel, about 5500 ft. above sea-level. At this point we descended the eastern slope of the Continental Divide, and passed by placer mines that have made many a heart to sing for joy, and many another to break from sorrow, until about 4 P.M. we reached the city of Helena, a view of which is given in Fig. 119. There are here some 15,000 people, and near by is the famous " Last Chance Gulch," which has yielded about 10,000,000 dols. in gold. It has been the hope of many people that they will some day walk the "Golden Streets," but in Helena, despite the name, this hope comes pretty near realisation, for you may here walk the silver streets, since, just before our visit, an excavation in the principal thoroughfare showed a well-defined vein of silver extending along it, and the burning question agitating the city throughout at the time of our visit was, to whom this vein belonged. At this place the benefits of our method of travelling were shown. We had been living on our train for several days, and in order to break the journey and give us a rest, this stop was made, and the entire party taken to the Hotel Broadwater by means of an electric car, a distance of three miles from the station. This was a delightful surprise, and gave us all a pleasant feeling of obligation to Raymond and Whitcomb for their fore-

z

Fig. 116. Spokane Falls.

Fig. 118. East Entrance to House Plains, Montana.

Fig. 117. Deer Park.

Fig. 119. Helena, Montana.

thought. Not only was the hotel a large and beautiful house, with attractive grounds, and a fine table, but adjacent to it, in the park, was a very large and luxurious natatorium, containing a large bath, fed from hot springs, into which the party plunged, and emerged thoroughly refreshed and ready for the dinner, which was worthy of the place in every respect, and to which our sharpened appetites enabled us to do ample justice. There was a " hop " at the hotel that evening, and at the proper time the party returned to their cars and woke up the next morning at Cinnabar, the entrance to the Yellowstone Park ; but how changed everything was to the writer, who had ridden past this very spot in 1883, when there was no railroad nor any house to be seen. The principal feature of Cinnabar is the mountain known as the Devil's Slide, and, judging from its red colour, no doubt his Satanic Majesty wore out not alone the seat of his trousers, but also a part of his skin. His presence here is still defined, for almost every house is a liquor-saloon ; but then in this section of country even the air is dry, which may account for the fact. At this point we took stages for the hotel at the Mammoth Hot Springs, a drive of about six miles. Now we were within a stage drive of Yellowstone Park, a map of which is given in Fig. 120.

As soon as breakfast was over, we hurried out and were soon seated ; the driver cracked his whip, the four horses darted forward, and we spun away over the level sandy road.

" It is exactly as I expected it to be," said someone beside us, and we agreed with him. There are scenes which burst upon you with over-whelming novelty, and there are others which exactly satisfy your pre-conceived idea. This was the outskirts of the Yellowstone Park, as we had so often fancied it ; the wild, almost treeless, waste of low hills, covered with pale green sage bushes, here and there a lonely white house, and four or five reckless horsemen dashing away before us, controlling big herds of scattered cattle. On the slopes were the most exquisite wild flowers ; one especially beautiful blossom was shaped like a pond lily, and bore on the same plant both white and shell-pink flowers. Little gophers, which the driver called " whang-doodles," scudded past us, and " cotton-tail " rabbits bobbed through the sedge. Once we passed a white prairie schooner lumbering slowly along, and, peeping inside, we saw a comely brown-faced woman and four or five black-eyed children. A little beyond was their discarded camp, and near it we noticed steam ascending from a pool of water. A boiling spring ! It gave us a strange feeling of the

FIG. 120. PLAN OF THE YELLOWSTONE PARK.

thinness and instability of the very ground, and it seemed as though we must be over a boiling lake. A few trees were scattered on the hillside, and soon we approached a fringe of verdure along the banks of the foaming Gardiner River.

At about noon we saw between the green hills, the smoking terraces of the Mammoth Hot Springs. There is something so unnatural in this strange great formation that its effect is almost saddening, and the wonderful beauty of its varied colouring only makes it the more unearthly. It is far more like a scene from the "Inferno" than a part of the United States, and it requires but little imagination to people this blasted mountain with throngs of demons. After lunch we began the ascent, first passing the Liberty Cap, which is an extinct geyser cone 52 ft. high and 20 ft. in diameter at its base.

We had all brought blue or smoked glasses with us, but, very fortunately, the sun was partially overclouded, which prevented the glare on the white formation, while occasional bursts of sunshine showed us the colours. The ascent of the Mammoth Hot Springs is tedious and tiring, and it requires constant care to avoid stepping in the boiling water which trickles over the formation.

Our first stopping-place was Minerva Terrace (Fig. 121), which is 40 ft. high, and covers nearly three-quarters of an acre. Its pillar-like furrows shade from white and delicate cream to deep red and orange, and the steaming pool of water is vivid blue. The spring of Jupiter Terrace, which we next visited, is the largest, being nearly 100 ft. in diameter, and its terraces are in many fantastic forms. It would be useless to attempt a detailed description of the wonderful springs, yet each one is slightly different from all the others. In some the colours vary from pale pink to deep rose-colour; while another is as white as snow, its crystal water covered at the edges with floating flakes like marble. The smell of sulphur from the clouds of steam is very perceptible, but not at all disagreeable. If objects are placed in these pools they are heavily coated with white soda in a few days.

Distant growls of thunder hurried us back to the hotel, which we had hardly reached when the storm broke. It lasted only a short time, and when the sun again burst forth, the high pale terraces gleaming in the light, and the columns of white steam rising steadily in the windless air, were clear cut against the purple clouds. The Mammoth Hot Springs Hotel is a large and well-appointed building, and it was with feelings of

Fig. 121. Minerva Terrace.

FIG. 122. THE DEVIL'S PUNCH BOWL.

regret that we left the next morning. We were going into a wild and unknown country, and however beautiful might be the scenery, the happiness of humanity unfortunately depends largely upon cooks, as Owen Meredith has so plainly told us.

Our morning's journey was a pleasant and an interesting one ; our road wound along the dizzy edge of a ravine, where great engineering skill had been employed in its construction. The long drive in the fresh air had given us famous appetites, but when we reached Norris's, our lunch-place, the scene which met our eyes was far from reassuring. The hotel had been burnt to the ground, and several tents supplied its place. However, we were pleasantly surprised by an excellent meal.

We then drove through the Norris Geyser Basin, which is a tract of blasted land, honeycombed with little geysers and boiling springs. As the traveller gradually approaches from above, the jets of steam rising above the low trees make it look as if he was nearing a collection of factories. All the afternoon we passed a succession of beautiful and wonderful scenes, and we often stopped to taste the waters of medicinal springs of iron, soda, and alum. Beside the road were many boiling springs, the most curious being the Devil's Punch Bowl (Fig. 122), where, it is said, he cooks eggs in the morning, and the driver informed us, as we gazed at the popping, snapping bubbles, that his Satanic Majesty was then at work. We could not see him, and from the odours which blew to us from the pool, we decided that we did not care to remain to breakfast.

Later we saw two beautiful lakes, whose sea-green waters gleam over a bottom of marble-white formation, and then we all alighted at the cliffs of Obsidian Glass, and picked up great pieces, which resembled coal until the dirt was removed, when they were exactly like very dark green glass. Obsidian is a species of lava, which, according to Pliny, was first found in Ethiopia. The name, however, seems to have been applied by the ancients to Chian marble, and is probably a false spelling of the Greek *opsianos*, signifying to reflect images, because the Chian marble was as hard to cut as volcanic glass, and was used for mirrors.

The Indians used this glass in making arrow-heads, weapons, and tools. Relics of this sort, which tourists find, seem to be made of the superior quality of Obsidian, which was procured at the cliffs. This cliff is 250 ft. in height, and is in vertical columns, pentagonal in form, but more or less irregular. It is the only thing of its kind in the world, and is much visited by geologists.

After passing the foaming rapids and fan-like falls of the Virginia Cascades, we reached the Cañon Hotel, and a little later started for a walk through the woods. Not knowing exactly where the path was, we thought we could climb down the wooded sides of the cañon to the top of the Great Falls. We did it, but would not advise anyone to repeat the experiment. Clinging from tree to tree, with the fearful abyss beneath us, we slowly slipped downwards until we reached the great rock above the fall, whence

FIG. 123. THE GREAT FALLS OF THE YELLOWSTONE.

we could see up and down the cañon (Fig. 123). Someone has asked us what we thought as we gazed down this valley for the first time. We did not think—in the presence of its surpassing beauty, every sense seems concentrated in sight. Never in our lives had we seen anything which could at all be compared to it, nor a picture which could in any measure do it justice. It is not alone its enormous depth, nor the fact that no human being has ever descended those precipitous sides; it is its wonderful and gorgeous colouring, which renders the beholder speechless and spellbound; 800 ft. in height, it rears its jagged towers of pale gold, with here and

2 A

there a turret or boulder of brilliant yellow or vivid scarlet, while the
sliding sands are streaks of violet and brown and rose, and at its foot the
river rushes in swirling eddies of deep green water lined with foam,
while just beneath the Lower Fall were two great snow-banks yet
unmelted. Instead of the greens and browns which Nature uses in the
painting of her other valleys and cañons, she here revels in gorgeous
colouring, like the old masters of art. And, as ever, at her highest
grandeur and beauty she is inimitable ; of course the most carefully-
coloured photographs of this cañon are gaudy and unnatural-looking. The
whole valley echoes with the sound of the Great Falls ; above us the
Upper Fall plunged on the rocks, and rose again almost half its height in
shooting towers of foam and floating mists, while just beneath us, the
Great or Lower Fall poured its roaring tide over a precipice 360 ft. high.
For many years there was no photograph of this greatest fall, except the
one from which Fig. 123 was prepared, which was taken from a ledge
three-quarters of a mile away. This, of course, does not at all do it justice,
and photographers and artists have sought in vain all along the banks for
a spot to which they might descend. A friend of the writers', one of
the earliest visitors to the park, once made the attempt with a party of
eager men. They clung to the rocks, and succeeded in going down a short
distance, when, to their horror, they found that the treacherous sand was
falling under them. A fearful struggle followed, and after some moments
of agony, they gained a secure footing and relinquished the attempt. But
the enthusiastic artist, Mr. Haynes, would not be thus daunted, and not
long ago he went to the brink of the cliff, and, having been lowered by
ropes, actually succeeded in securing a photograph. It is told that when,
on the return of the intrepid man, the picture was developed, as the
magnificent outlines emerged from the dark glass, the anxious crowd of
watchers burst into wild applause.

The next morning we procured horses and rode through the pine
woods, visiting both falls, and keeping as much as possible near to the
edge of the cliff. It was a perfect day ; the scent of the pines was
delicious, and all along the path were violets, spring beauties, and others
of our flowers of early spring. And, indeed, the air was like that of an
April day, and beside the road were heaps of snow. In a mossy valley we
saw a beautiful little fawn which had just died on the brink of a pool.

In the afternoon we went again to the Rapids of the Yellowstone, and
the Upper Fall, and found it even more beautiful than before ; the sunlight

glinted through the trees on the wavering mist, and in it quivered the colours of a rainbow. We watched the water roll over the precipice in ragged ropes of foam, until the sun sank behind the cliff, and the floating mist above the arrows of the rising spray, was tinged a soft pale rose.

On our return we saw buffalo tracks in the mud, and once our horses shied violently ; one, having jumped quite across the road, stood trembling all over. We gazed into the woods on every side, and seeing nothing, said, " It might have been a bear !" and the stories we heard next morning decidedly confirmed us in this supposition, for, when the writer was mounted on high beside the driver, he at once began :

" Did you hear what happened last night ? Two of the drivers had to go in the barn to sleep, and as they entered they thought they heard something snorting, and there, in one of the stalls, was a big bear. They ran like mad, and he after 'em. One fellow got out of the yard gate by luck, but 'twas so dark the other couldn't find it, and he was that wild, he just tore down the light plank fence to get away." Our road now lay through the pine woods, and along stretches of plain covered with sage bush and beautiful wild flowers. We passed many boiling springs, and one little geyser shot a fine shower of water into the air. After another lunch in the tents at Norris's, we wound along by the Gibson River, and saw its beautiful white fall spreading over the rocks.

We now drew near our destination, and were just clattering over a little bridge, when the driver called our attention to a great lump of soda formation on the opposite bank, saying :

" That is the Riverside."

" How it hops !" remarked some one, watching the boiling water leap up in the central hole, " Does it always do like that ? "

" O, yes, it often——"

" Wait ! Wait !" we cried.

He pulled up the horses, and even as we spoke the geyser roared and cast up a thin wavering column of foam. Higher and higher it rose, until it spanned the river with a streaming arch of water. We watched it, speechless, until it became quiet. Who can ever forget the sight of his first geyser ?

We had now reached the Upper Geyser Basin, a strange and blasted region, crusted with soda, and sprinkled with the mounds of geysers and boiling springs. As we passed the curiously shaped cone of the Bee Hive, the driver again stopped the horses, and exclaimed, " Well, you people are in luck!" and in about a minute this geyser also rose nearly 21 ft. high in the air.   Across the level beyond we could see our destination, the Paper House.   This hotel of the Upper Geyser Basin was burned last year, and a structure, principally composed of laths and brown paper, was erected in haste.   This is not the place for a conspiracy to be laid, for almost every sound can be heard in all directions.   About half a mile from it we passed the enormous cone of the Castle, one of the largest geysers (Fig. 124).

FIG. 124.   THE CASTLE GEYSER FORMATION, UPPER BASIN.

The imposing crater of this geyser is 100 ft. in length by 75 ft. in width, and its column of water is from 30 ft. to 50 ft. in height.   Sometimes, when it has husbanded its strength an unusual time, it will cast a jet 100 ft. high.   The little indicator at the left of the geyser was leaping at a great rate, and several soldiers and tourists were standing near.   We called from the carriage :

" Is it going off ? "

" It's twenty-four hours overdue now," was the answer.

We decided not to wait, however, and walked out upon the glaring white waste. Such a scene as this is almost indescribable; hot water trickled over the crust on every side, great pools surrounded us, into which we could look down incalculable depths into the clear water and see the varied gorgeous shades of blue and green. A particularly lovely one, which is called the Morning Glory, is shaped exactly like that flower, and is of delicate tints of pale amber, brown, and faint pink. One curious geyser, called the Sawmill, was continually making the sound of a saw with two jets of water snapping up and down. It seems unsafe to walk on the white crust with the roar and rumble of turbulent water about and beneath you. We returned to Old Faithful, which is just back of the hotel, at six o'clock, and were in time to see it shoot up a great pillar of steam and foam against the deep blue sky, while the span of a rainbow shone in the showers of its glittering spray.

Immediately after supper we betook ourselves to the Castle, which had not yet had its eruption, and there remained, fighting mosquitoes, and when they became unendurable, standing where the sulphurous steam blew from some pools. The sunset glory faded from the little fiery clouds blowing across the western sky, and there shone the arc of the new moon, and still we stayed. It was impossible to help personifying the geyser, for it acted like an angry demon. It roared and spluttered, and the explosion seemed imminent. At length we left it reluctantly, and when we rose next morning were told, to our joy, that it was still preparing for the eruption. We waited until noon, when we decided that we must see the prismatic lakes, though we all agreed that "that beast was just mean enough to go off while we were away." As we passed it, jets of water spurted from the crater several feet high, but, as it had done this before, we proceeded. We were looking out of the back of the carriage when clouds of steam appeared over the trees, half a mile away. We turned our horses and galloped them back, reaching the Castle the instant before the eruption. We had hardly arrived, when, with an awful subterranean roar, the geyser cast a pillar of foam and steam 150 ft. into the air.

We then drove to the Devil's Punch Bowl, the many-coloured formation of which is shaped much like a bowl, and is brimming with boiling water, which ripples over a fathomless hole of cerulean blue. Leaving our carriage, we walked some distance across the white crust to the wonderful prismatic lakes. It is a curious sight, this wide area of boiling water, from which clouds of sulphurous steam ascend, and the exquisite beauty of the blending colours is overwhelming. Each lake is slightly different from all the others, but in order that the reader may be better able to picture to himself their appearance, the writers noted exactly the sequence of colour in one. It was pale salmon at the edge, which shaded into brown, thence it glowed to yellow, fading gradually to pure white, while the bubbling centre was deep blue-green.

FIG. 125.   THE GRAND GEYSER.

On our return to the geyser basin we visited the Grand, and obtained photographs of the cone (Fig. 125), which is a very large one, and also of the water in the cone before eruption, and of the geyser in action. We also saw the Splendid (Fig. 126).

For the convenience of the reader the following Table of the geysers of this basin is presented :

Name of Geyser.	Interval or Period.	Duration of Eruption.	Height of Column in Ft.
1. Old Faithful	55 to 70 min.	3 to 5 min.	100 to 150
2. Bee Hive	7 to 25 hours	3 to 18 min.	170 to 219
3. Lioness	Not known	About 3 min.	60
4. Lion ...	do.	About 5 min.	75
5. Giantess	14 days	12 hours	250
6. Saw Mill... ...	Very frequent	1¾ to 3 hours	15 to 20
7. Grand ...	16 to 31 hours	10 to 42 min.	95 to 200
8. Turban ...	About 15 min.	15 sec. to 5 min.	25
9. Castle	Once in 48 hours	30 min.	100
10. Giant	Once in 4 days	1½ hours to 3 hours	130 to over 200
11. Young Faithful...	Very frequent	...	10 to 30
12. Oblong ...	Once or twice daily	6 min.	
13. Splendid ...	About 3 hours	4 to 10 min.	200
14. Grotto ...	Several times a day	Half-hour	20 to 60
15. Fan ...	Three times daily	5 to 9 min.	About 60
16. Riverside	Three times daily	10 to 13 min.	About 60

These geysers were once quite regular, but before the Park belonged to the Government, tourists were in the habit of "soaping" them—that is, throwing soap down the crater, in order to produce an eruption. This disturbed the regular action of the geyser, and the Government now prevents it by making it a penal offence, and guarding the geysers. The landlord of the hotel took us to visit his laundry, which is a very novel one, being merely a large tent beside a great boiling spring. We were told that the first person to try this experiment was a Chinaman, who was so unfortunate as to choose a geyser cone then quiet, but not extinct. The result of the soap may be easily imagined. One morning the geyser rose early, and Chinaman, tent, and "washee" were knocked sky high.

We left the Upper Geyser Basin in the afternoon, and our drive was enlivened by the fording of a swollen stream where the water came up to the bottom of the wagon. We all alighted at the strange tract of land which has been so fittingly named "Hell's Half Acre," and looked into the enormous crater of the Excelsior Geyser. This immense pit of irregular outline is 330 ft. in length and 200 ft. wide at the widest part. This geyser has eruptions every fourth year. The first record of this most powerful of all geysers is as follows :

" Colonel Norris asserts that he at first heard its spoutings at a point six miles distant, but reached the scene too late to witness them, although he saw the effects of the eruption upon the Firehole River, which was so swollen by the flood as to wash away some bridges over the small streams below. In February, 1880, the Excelsior became frightfully violent in its eruptions, causing the earth to rumble, and filling the valley with dense vapour. The period of action began about 10 P.M., gradually becoming later every night, until on the 1st of July, the eruption took place at 10 A.M., showing a loss of 12 hours during nine months. Colonel Norris reported that in the summer of 1880 the power of the eruptions was

Fig. 126. The Splendid Geyser.

almost incredible, 'elevating sufficient water to heights of from 100 ft. to 300 ft. to render the Firehole River, here nearly 100 yards wide, a foaming torrent of steaming hot water, and hurling rocks of from 1 lb. to 100 lb. in weight, like those from an exploded mine, over surrounding acres.' The Excelsior has increased in activity ever since, giving two or more displays daily, sending out a compact body of water from 60 ft. to 75 ft. in diameter to a height varying from 60 ft. to 300 ft. It is a sufficiently awe-inspiring experience, as the writer can affirm, to stand at the verge of this steaming lake, upon the hollow crust which projects over the boiling water, and peer down upon the agitated surface as the clouds of scalding vapour are occasionally lifted by the breeze. But when this geyser is in action, the awful noise and concussion produced by the falling water, accompanied by rumblings and vibrations like those of an earthquake, and the disagreeable habit of vomiting up stones, which is a special characteristic, warrant the visitor in keeping a safe distance away during the display of its terrible power."

We reached the Lower Geyser Basin in the afternoon, and immediately went upon the white formation, and after passing many little geysers and many black holes, from which the accumulated steam burst from time to time with a dull thud, we climbed a little hill and saw before us the mud geysers which have been so appropriately named the Mammoth Paint Pots. This cauldron of bubbling mud measures 40 ft. by 60 ft., and its rim is from 4 ft. to 5 ft. in height. It is filled with a mass of pasty clay, which boils sluggishly, and spits from its bursting bubbles, globular masses, cones, rings, and jets of mud. Some parts of it are so hard that little paths have been worn upon the baked crust, and it is decidedly startling when you are walking by a quiet lump of mud to have it suddenly explode and cast a soft bullet close beside your face. The main basin is white, but the mud puffs or cones scattered about are delicate shades of old rose, and all look exactly like boiling paint.

In the evening a party of us assembled on the formation to see the Fountain's eruption, which was momentarily expected. In front of us was a great broad spring, in whose centre were rising bubbles. One of the ladies and a small dog were seated by the edge of this pool looking into its clear sapphire deeps. Suddenly, without any warning, the centre of this lake rose with a fearful roar. The lady fled, and the dog galloped over the formation, yelping with terror. Unlike any other geyser we had seen, this beautiful Fountain rushes into the air, not from a crater, but from the centre of a pool, rising vertically to the height of 60 ft. to 80 ft. in many jets, and with other streams shooting out in all directions.

We returned next day to the Mammoth Hot Springs, whence the tourist may make a day's excursion to the wonderful Petrified Forest. This is a very paradise for geologists, for the strata contains all sorts of animal and vegetable remains, and all over the slope of the hillside are the great trunks and branches of the trees petrified into clear agate or opalised, with the cavities full of quartz and calcite crystals. As you stand in this

2 B

frozen forest it is easy to fancy you are on the towering mountains of the "Arabian Nights," where the princely seekers for the wonderful bird were turned to stone.

We reached Cinnabar the next day, just in time to escape a heavy storm. The thunder crashed about us as we hurried from the stages, and as we boarded our train the rain descended suddenly like a sheet of steel.

Having said farewell to all the wonders of the Yellowstone in a terrific thunderstorm, seemingly an appropriate method of leave-taking, we were soon speeding along our homeward journey. The continued rains had been productive of frequent "washouts," which delayed our progress somewhat, but the various places were so interesting, that delays produced no visible effect on the spirits of the party, and after two days' continuous travel we reached Minneapolis in good order and in proper shape to celebrate the great National Holiday.

It seems hardly fair to impose on a narrator the delicate task of weighing rival claims, and having, as we hoped, safely balanced Tacoma and Seattle, the writers believed they had steered safely between Scylla and Charybdis; but these self-congratulations were dispelled at this stage of the journey. Seattle and Tacoma are separated by 35 miles, but Minneapolis and St. Paul are but six miles apart. Admiring, as all Americans must, these four great cities, the writers intend to endeavour to preserve the balance of power in this case as in the former, and should they seem to lean towards either city, it is only by accident in circumstances, and not by any intention.

We reached the charming city of Minneapolis at 11 A.M. and went at once to our hotel, where our baggage was delivered within an hour. Without making any invidious distinctions, we would like to say right here that we have never stayed at a nicer hotel than at the West House, an excellent picture of which is given in Fig. 127. Coming as we did from a journey of several days, the quiet of this beautiful house seemed to soothe us and put us in a frame of mind to enjoy all the pleasures of this lovely city. At this time the Exposition Building had considerable interest to us, for the great National Republican Convention had just finished its sessions, and renominated as their candidate the present President of the United States. As Minneapolis takes into its city limits Nicollet Island, it is natural that the citizens should seek to construct a first-class bridge for the great amount of travel passing to and fro—and this is not only a fine structure, but is also quite a feature in the landscape. The steel arch is an

FIG. 127. WEST HOTEL, MINNEAPOLIS.

FIG. 128. THE BOSTON BLOCK. MINNEAPOLIS.

object of very great beauty. A short distance above the hotel is the building of the Young Men's Christian Association. As education has always kept pace in the United States with the gradual settlement of our outlying districts, so in Minneapolis, although a thoroughly developed city, we find this trait has not been overlooked, and their Public Library is a structure that any city in the world might be proud of.

The enterprise of the citizens, and their desire to erect beautiful and substantial structures, find, as usual, a response in the General Government, and the United States Post Office is an evidence of the appreciation in which the National Government holds this city. Behind it is probably the finest building in the city, that of the Guaranty Loan. There is a fine restaurant in it, and the whole finish is superb, marble wainscoting being prevalent in every floor. The public are admitted to the roof for a fee of 10 cents, and no one ought to neglect this, for the view is simply delightful, and the vista opened to the spectator is well worth all the trouble. (See Fig. 128.)

The building is but a short distance from the West House, some three minutes' walk, and a better idea of the city and its environs cannot be had than from this point. Being at the head of navigation, the lumber interest is one of the most prominent in the city, shipping 343,000,000 ft. in 1890, and, naturally, the building in which this centres would be worthy of it, as the reader may easily judge from the general style of architecture in the city. There are many other fine buildings to be seen—in fact, so many that a selection is quite difficult, but the two shown in Figs. 128 and 129 are a fair sample, and as such are offered to the reader, the Masonic Temple being offered to all craftsmen. Although this city and St. Paul are rivals in trade, yet they are connected by a trolley electric railway, and the cars passing in both directions were full to overflowing with the people in transit.

By a little diplomacy and a detour of a few blocks, our party obtained seats, and were speeding off to the park, some two miles away, on the route shown in Fig. 130.

This park is a charming place, with lovely walks, beautiful trees, dells and cascades. If it contained nothing else but the sheet of water shown in Fig. 131, it would be famous indeed, for these falls are the ones immortalised by the Poet Laureate of America in "Hiawatha," and we realised, as we stood face to face with them, that this was Minnehaha (Laughing Water).

We commenced to feel extremely sentimental, and to reflect in the train, " As unto the bow the cord is, so unto the man is woman ; though she leads him, she obeys him ; though she draws him, yet she follows," when our sentiments were rudely dispelled by the lozenge-boy's " Peppermints and winter greens, two for five," and we rushed from our car to be

FIG. 130. NICOLLET AVENUE.　　　FIG. 129. GLOBE BLOCK.

landed at the Hotel Ryan, in St. Paul, in time for a good dinner and plenty of noise as a concomitant, for the city was after a Fourth of July celebration, and, when St. Paul starts for a thing, she usually gets there. Ordinary means failed to express the feelings of these enthusiasts, and chlorate of potash was placed in lines on the rails of the street cars. The noise was something stupendous, and although one of the writers served 100 lb. and 200 lb. guns during our "late unpleasantness," he

FIG. 131.   MINNEHAHA FALLS.

FIG. 132.   MINNESOTA STATE CAPITOL.

must own to something in excess of the entire concussion of his whole
battery.

But St Paul is a great city, full of enterprise, with many fine
buildings, and producing a fine-looking set of citizens. If anyone doubts
this, let him look at their book entitled "The Resources of St. Paul."
The place was settled in 1838, and in 1841 there was a Jesuit log chapel
here. In 1849 it was surveyed, and in 1858 Minnesota was admitted to
the Union, and St. Paul became its capital. Following the usual custom
of western cities, it commenced to annex the surrounding country, and from

FIG. 133. HIGH SCHOOL, ST. PAUL.

90 acres in 1849 it grew to 2561 acres in 1854, and thence to 35,482 acres.
If Minneapolis had not started up and commenced to develop in the same
line, St. Paul might have annexed the balance of the State by this time.
The present population is upwards of 80,000. The State Capitol (shown
in Fig. 132) is a fine building, located in the centre of the city and of
imposing appearance. The city, being largely settled from the east, has
given much attention to education, and many fine schools have been
established. The High School building seen in Fig. 133 is a structure any
city might be proud of; indeed, there are few towns which can show a finer
structure for this purpose.

After the "glorious Fourth" had passed away and the city assumed
its normal condition, we took a drive around it and the suburbs, visiting
Bear Lake and other watering-places with which St. Paul is liberally

provided, for there are 10,000 lakes in the State of Minnesota, which contains the head-waters of the great Mississippi River. Summit-avenue is the fashionable location for the " 400 " ; it overlooks the whole city, and is a most beautiful street, as the reader will readily believe when he looks at Fig. 135, which shows a little of this favoured spot. The thermometer

FIG. 134. SCENE IN MINNESOTA.

FIG. 135. SUMMIT AVENUE, ST. PAUL.

in winter will go to 40 deg. below zero in St Paul, but the inhabitants always tell you the air is so dry they never notice it. Perhaps not, but the narrators don't incline to that idea, as they have tried 20 deg. below and don't like it, dry or not. Theoretically it sounds well, but we remarked the universal use of furs, and think 20 deg. above, preferable every time ;

still the St. Paulites look well and seem healthy. We enjoyed our short stay there, and can heartily recommend the Hotel Ryan to anyone who likes good rooms, good meals, and good service. The next night we were speeding to Niagara in the vain endeavour to escape the Christian Endeavour Society. But they were everywhere—in trains and by the wayside ; at all points along the road, and at the stations. Still, our conductors managed to arrange our meals satisfactorily, for the name of Raymond and Whitcomb was a power, and, after a day at the Falls, in which the writers visited the tunnel under construction by our old friend Mr. Clement, we took the night train for New York, dodging two or three Christian Endeavour crowds, who were *en route* to their great meeting at Madison-square Garden. These people are to be commended for their work, and just so long as they stick to it, should be encouraged ; but when they take a sanctimonious attitude and talk about the iniquity of keeping the Columbian Exposition open on Sunday, and propose to stay away, &c., they go too far, and their inconsistency is shown by their patronage of railways which operate Sunday trains, and by their meeting in a garden which always has Sunday concerts on the very spot they use. Still this is neither the time nor the place to discuss theology or morals. We ended at New York the most delightful trip possible, the interest of which lasted from the start till the end of the journey, and separated to our everyday occupations with mutual regrets.

As a contrast to this recent and most comfortable excursion over the National Park, the story of a visit made by one of the writers seven years before, on the occasion of a summer meeting in the vicinity of the American Society of Engineers, may be introduced here, not only because it describes this wonderful region in greater detail, but also because it will afford some indication of the change and development that have taken place in a short time.

After a most delightful reception by the railways of the north-west and an equally delightful convention at Lake Minnetonka, in Minnesota, a number of the Civil Engineers who had just concluded attendance on the meeting, decided to accept the hospitable offer of the Northern Pacific Railway and to visit the Yellowstone Park ; and let all fatalists and superstitious persons generally take note that the party numbered thirteen, and that all are alive at this writing. The outfit was purchased in part at

2 c

St. Paul, and much amusement was experienced by the purchasers. Being novices at the business we bought everything recommended; in several instances we purchased too much, and in one instance—one only—too little. This last omission was an important one, it was blankets. A double blanket seemed ample, indeed almost unnecessary, and we were beguiled by some experienced (?) travellers into ideas of how soft the earth was, and how careful we must be not to over-burden ourselves. The writer's army life might have served him a good turn, but for its peculiar phases. First, it was in a warm climate and two blankets then were ample, and next it was principally on the coast and in the sand, which was soft. However, we only took a double blanket each, and many a time the experienced traveller received the benedictions of his frozen and aching friends. For the earth is one of the hardest beds encountered, and if she is the common mother of us all, certainly has a very uncomfortable lap to hold her progeny in. We started from St. Paul on a lovely June day in the most exuberant spirits, each man expecting to bring home a grizzly at least. Equipment is a matter which Englishmen carry into details which astonishes an American, and yet the truth lies between practice of the two nations. We are very apt to hastily prepare ourselves, while an Englishman may insist on carrying his bath tub with him. We passed rapidly through the famous wheatfields of the Red River Valley of Dacotah—the name valley being a misnomer, as it is really a plain and produces 25,000,000 bushels per annum—and marked the peculiarities of construction of the North Pacific Railway. One can never pronounce with certainty on the action of another man's mind, but as engineers it did occur to us that the sinuosities of the railway seemed rather unnecessary on pretty flat lands and not inhabited to any extent; of course we did not mean to say the liberal land grant of our liberal Government per mile of construction, influenced the location, although when we reached the so-called " Bad Lands" it is true the line seemed more direct.

Later on the railroad crossed the river and we passed from Morehead, Minnesota, to Fargo, Dacotah. We reached the latter place in the night, and the writer, roused by the stoppage of the train, drew up the curtains of the sleeping berth and saw in this western territory a mast at least 200 ft. high, and an electrical installation which is finer than the one in the centre of New York City. On our return trip, passing Fargo in daylight, we inspected it, and could hardly realise that one year before our visit, it had been a railroad town on wheels, and contained only the working crews of the constructive force of the N. P. Railroad. Now it had a brick hotel many brick stores, an opera house, telephones, electric lights, and all the accessories of modern civilisation, including drinking saloons, gambling houses, &c.

West of Fargo is one of the richest districts of the north-west, Cass County by name. It has produced 5,000,000 bushels of wheat in a year and all the land is taken up by settlers. At Dalrymple, 18 miles west of Fargo, is the wheat farm of Mr. Oliver Dalrymple, which occupies 117 square miles. The train continued to run through wheatfields for hours, and finally we reached Bismarck.

West of Mandan, we entered the Bad Lands. The first prairie-dog village was sighted here, with this peculiar little animal, resembling a huge rat, seated on its hind legs at the entrance to its burrow. The usual results followed ; the passengers fired shots from their revolvers, and the animal escaped unharmed, giving the short, defiant bark from which its name is taken. Although the name " Bad Lands" is given, yet there are flowers to be seen in great profusion, and altogether the country is not un-attractive. The buttes or hills, which rise on all sides, are most peculiar. They resemble huge castles, and might seem to belong to the middle ages. The colours are also most remarkable, being black, blue, and grey. They are evidently of volcanic origin, and formed partly by the action of water upon clay. In some instances they are 300 ft. high. The summits in some cases emit smoke, and this keeps up the illusion of an inhabited castle. As our train went through this section by the light of a beautiful moon, the weird effect was most realistic, and we could almost imagine we heard the tread of the warder, and were in expectancy of a challenge or a shot from some frowning battlement. The clay seems to have been solidified by pressure, and the heat beneath turns it into terra-cotta, while the rain cuts the surface into all sorts of strange shapes. The coal crops out in spots, and is mined for the use of the railroad ; it appears to be of good quality, although containing much sulphur. The name of " Bad Lands " is supposed to have come from a phrase of the French voyageurs who stated the country to consist of *terres mauvaises pour traverser.* Some miles further west we crossed the Little Missouri river, and from it the station takes its name.

The Marquis de Mores has established here the head-quarters of an extensive stock-raising enterprise, and is shipping dressed beef to New York City. It is singular how all stock in this country is beef. A stock raiser of a few thousand head of beef looks down on a man with ten times the number of sheep. "Nothing but a sheep-raiser" is a term of contempt in Montana. It was in this region, and on some distant buttes, that the eagle eye of a fellow-traveller detected antelope. At first we could see nothing, but by means of a glass we saw what he had pronounced on at a glance by his unassisted but well-trained vision.

At little Missouri we saw a gathering around the baggage car, and on inquiry from a cowboy were told it was "only a stiff." We found it was a man who had undertaken to "run the town." He and a companion had come in and taken possession of a saloon, and after dispensing liquor freely to a crowd, had started out to shoot the Marquis de Mores. Unfortunately for their success, the Marquis fired too, and the result was the "stiff," who was taken in our train to Billings to be "planted." At Billings, also a growing town (it absolutely grew during the two weeks we were in the Yellowstone Park), attempts at irrigation had been made, and we passed ditches near the railroad, and branching into the prairie for many miles. Finally, about 1 P.M., the party disembarked at Livingstone in good order. Livingstone at that time was the nearest point to the park (being some 66 miles distant). A branch road has since been constructed, but at the time we made an entrée, it was only on paper. We held, at once, a council of war and decided to divide up the work. One party inspected transportation, another arranged for "quarters in the field," and the third was the "commissary of subsistence." In this country all negotiations begin and end with a drink, hence the committee had to be men who could, to use the phraseology of the land, "stand the racket." The head-quarters committee obtained a large tent which had belonged to a Government hospital, and could easily accommodate thirty men. The commissary party were equally successful, and had an abundance of provisions, indeed a superabundance, as we brought back a large quantity, but the transportation committee could only report "progress." In truth, they were so divided in their own minds as to what they wanted, it was little wonder they failed to settle on anything; moreover, each native had a new method to propose, new advice to give, and new horses to offer. Why, the stories told to this committee would have made old Ananias blush with shame; he was a tyro to these frontier men, and even if we had heeded one-tenth the advice given us about being cautious in our dealings with this, that, and the other man, we should have been convinced that Livingstone was even worse than the cave in Ali Baba, for that only held forty thieves. At last, after thirteen saddle horses had been engaged and all countermanded, it was decided that each one should get his own horse or horses, leaving the commissary department to arrange for its own teams and methods. Accordingly we divided up into groups, and two of us who had decided to go in the saddle, each secured good Indian ponies, called by the natives "Cayouses" (Ky-use). Two other saddle horses were obtained as a change for the balance of the party ; but they decided to go in wagons and on buckboards, while the provisions and the tent went in two army wagons. These arrangements occupied the balance of the day, and we retired, prepared to start at daybreak. That is, we were prepared in our minds to start at this time; however, a preparation is not always an accomplishment, for the teamsters and the cook, and horses, and one thing after another, and the farewell drinks, and the sobering up of the guide, &c., prolonged matters till 8 A.M.; in fact, it took nearly one hour to properly swear the teams into their respective wagons. The teams, it must be understood, were not "sworn in" as the United States swears in its soldiers ; it was a much more cursory method. The writer and his friend, who were the only two properly mounted, rode on in advance. Livingstone stands on a plain, and some six miles distance the gorge commences, through which we wound for some distance. In passing through this gate of the hills the road winds along the Yellowstone River, and on the opposite side appears Bear Gulch. The air was so pure it seemed but a mile off to the hills which formed the entrance to this gorge, and after riding for two miles it seemed just as far off. This rarity of the atmosphere is a remarkable phenomenon, and causes many amusing mistakes. It is said that an Englishman started to take a "bit of a constitutional" before breakfast, intending to go to the foot of a hill he thought about three miles off, and was found in the afternoon, having walked over 15 miles without reaching it, trying to jump over a brook about 1 ft. wide ; he explained this peculiar action by saying he had been so deceived by the distance of the mountain, that he was ready to believe this brook was 10 ft. wide instead of one. In due time we also reached a brook, and found a wagon had conveniently broken down and broken the bridge also. The writer was told by some men who were trying to unload the wagon, that the ford above was "all right." He saw a twinkle in their eyes, and knew they had set him down for a "tender-foot," so he took the precaution to cross his legs over the horse's neck, and it was well he did, for the water came up to the saddle. One of the wagons had to stop here, and a man was sent back for another team ; finally another wagon was procured, and all crossed at last by doubling the teams on the wagons. Beyond this brook was a wood, and here was a tent; at its entrance were two barrels supporting a rough board on which were a greasy pack of cards and a dice-box. Outside, marked on the side of the tent, was seen a " fiery untamed steed " on his hind legs pawing the air. This is a conventional sign, and means that in that place the celebrated game of "stud-horse poker " is played. The proprietor came out in a

FIG. 136. THE YELLOWSTONE FROM THE FIRST CAÑON.

pretty tough-looking suit, wearing a much tougher-looking visage, and said he would "shout," and after going through the formula of asking each man what he would drink, announced he only had whisky, of the 40 rod character.

The writer here met with an episode, if the lady will permit herself to be so called. While at this place along came a lady from Helena, Montana, escorted by the son of a livery stable keeper in Livingstone as a guide. She had resolved to "do" Yellowstone Park, and having been disappointed in her party, and having a limited vacation, being in charge of a young ladies' school, she had started alone. She was a typical Western although born in the East, was a good rider so far as sitting well in the saddle was concerned. Her "staying powers" proved later to be an uncertain quantity. The lady wore rather a coquettish costume, and had a revolver strapped around her waist. She declined the invitation of the country to "light and brace up," and rode on with her escort.

The writer concluded not to wait for the wagons to come up, and rode on later, and being well-mounted, soon overhauled the two, when racing was proposed and kept up at intervals till we came to a ranche where we were to dine. Although it was June still the day had not been over warm, but the sun had blazed on us for two hours previous, and we were glad to alight.

After dining on ham and eggs, the lady and the guide got up a pistol match ; your correspondent saw them shoot one round and promptly declined their invitation to come in. The lady put five bullets into a space about 3 in. square at the distance of 200 ft. The guide put in one or two. When I shoot I want a fair-sized target, a grizzly bear or two, but I want him to wait where he is while I reload, as it always takes at least one round to get my hand in.

Our country is a curious mixture of everything. Here, in this little house by the roadside on the border of civilisation, we found a parlour organ, and it served to pass several hours, as we had to wait till the cool of the day. One party passed while we were there, and they looked forlorn enough. We advised them to wait and cool off, but in their self-will and pride they declined, and we afterwards passed them still forlorn some five miles distant going at a snail's pace. We started about 3 P.M., just as the sun was slanting a little, and, riding alongside the river, arrived at another ranche about 7.30. Our party came in about nine. If there is one thing more aggravating than another, it is to come in hot and tired and to see people sitting comfortably in the shade nicely cooled off and well-fed. Opposite this ranche, known as Fridley's, Emigrant Peak may be seen.

The next morning we started in fine spirits ; the Helena lady got into her saddle with a jaunty air and started off with her guide. We followed more leisurely, and got our teams and baggage into shape, which took time.

About a mile out we met the guide coming back leading the horse with the side-saddle. All hands clustered round to hear of a mishap perchance.   But no, that reckless racing of the previous day had proved too much for the lady ; she declared she would ride no more, and arranged to go the balance of the trip on a buckboard.   Then they laid all the blame on the broad shoulders of this innocent writer.   This little incident was soon forgotten in the beautiful scenery we passed through.   The road wound along the banks of the Yellowstone River, which was broad and swift, while on the opposite side a range of peaks lifted their heads crowned with snow into the bluest of blue skies.   The variations in

Fig. 137.   Cinnabar Cliffs near the Yellowstone.

temperature are remarkable, and it requires care on the part of the tourist to avoid bodily discomfort.   In the early morn it is very cool, but after 10 a.m. it becomes scorching hot, and if the tourist is wise he will avoid riding between 11 a.m. and 2 p.m.   In evening it is again cool and the twilight lasts at this season late into the night.   The writer and his companion had ridden on in advance of the party for the purpose of having his horse shod.   He had cast three shoes, and the fourth was in bad shape. It was no easy task to find any one, but after one or two attempts, at a

Government forge belonging to the U.S. Engineers, a man was found who would undertake it. While we were still on this errand we overtook two men on horseback. Fastened to their saddles were blankets and camp utensils ; at each of their saddle bows was a Winchester rifle, and each man wore a large-brimmed hat with a leather band around the crown. One merited a particular description, since we saw a great deal of him afterwards.

He had a red, sun-burned face, his hair was cut close to his head, his face was badly scratched and one eye was black. He had a fine physique and a pleasant expression to his face, despite its disfigured appearance. We learned subsequently he had run against another man's fist the night before, but had laid out his man later in the fight. He had on a leather vest, fancy hunting shirt, and leather trousers fringed at the side. He had no coat, but around his waist was a belt full of cartridges for his Winchester, which was within easy reach, and he had a heavy breech-loader revolver and a hunting-knife in his belt. He was mounted on a cayeuse, and his spurs were of the Mexican kind, with a rowel about 2 in. in diameter, with jingles attached to it which tinkled as he rode.

To use the language of the country we " passed the time of day with them." My companion said afterwards, he was uncertain whether they would return our greeting or fire a bullet at us. Fortunately for us they chose the former method, and we passed quietly on, separating to go on the horse-shoeing business a few miles beyond. We now climbed industriously up the side of a pretty respectable hill, and here was offered a sample of the flexibility of the human will in shaping nature to its own ends. A gorge occurs here, shutting the river within narrow limits ; a narrow shelf of land comes out at the foot on one side of the river.

One " Yankee Jim," with the true instincts of his name, had built directly across the only route a large gate made of hemlock trees ; he claims the right from the United States Government to exact toll, and as he is a good shot and a clever man if not crossed, no one has thus far disputed his claim, or if they have done so none but Yankee Jim knows it or where the disputant is. He has alongside on the right-hand of his toll-gate a bar-room, and on the other side, his dwelling, where he entertains visitors, even keeping them over night if desired. The only route to his house is through his bar-room, although as stated, they are on opposite sides of the road. Still he won't permit any one to go to the house till they have been through the bar-room. He welcomed the party with

2 D

evident satisfaction, and promptly insisted on "shouting." This, of course, included all the present habitants of his shanty. The true inwardness of this hospitality will appear when the etiquette of the country is considered. It was incumbent on the visitors to "shout" in return, and then as there were six or seven of us to "shout" in sequence—all drinks and cigars being 25 cents each—Yankee Jim reaped some 10 dols. before dinner.

While waiting for this important meal to be prepared, our two friends before described rode up and were greeted by the host as old acquaintances. They "shouted," and then "shouting" was renewed all along the line.

At last dinner was ready, and while not elegantly served, it was very sustaining, quantity being the desideratum. To use a Westernism, " It was fillin' for the price." We rode till late that night, and then made our first camp. Not being experts, and not having engaged a guide, we chose a spot convenient to a pleasant stream of water. It afterwards proved to be equally convenient to a patch of prickly pear, a species of wild cactus, and we remembered this camp for days afterwards—indeed, it is not at all improbable that samples of the prickles are yet in various members of the party. We were all tired, but each one cheerfully went to work making tent-pegs, and we soon had the satisfaction of raising our canvas roof. After a meal of ham and bread and a cup of good coffee, each man filled his pipe and rolled into his blanket. Was ever ground so hard ? It really seemed as if it had been baked into clay or some semi-rocky formation. It was altogether too bony for "Mother Earth's " lap, and it must have been her spine. We had, after twisting and turning, gradually relapsed into a semi-somnolent state when we were all roused by the trampling of horses and a shout, from the lungs this time ; the writer ran out in his stocking feet and into some prickly pears. What he said was unnecessary, although quite emphatic. We found our two friends again, and pressed them to "light and shout." They came into our tent, and made themselves very agreeable for an hour, but declined to sleep under its shelter because they preferred the open air. Their "outfit " was quite complete, for they had four or five blankets each, and did not find the ground so hard. Just before withdrawing, the principal one whispered to me that if I and my companion would get up early, he would pilot us to Mammoth Springs ahead of the party and show us around. His offer was accepted, and we got under way the next morning by 5 A.M. We first interviewed a guide, my companion and myself having made up our minds to "have our head-quarters in the saddle," and leave the balance of the party, who were loaded down with camp impedimenta. This guide was a typical Montana man, and we were pleased with his quiet ways and soft accents. He invariably spoke with the rising inflection, and had a most musical voice. We learned afterwards he was very much dreaded among the lawless element, and had shot some sixteen men already, and yet "it seemed as though butter wouldn't melt in his mouth." It is with trembling hand these words are penned, for if he obtained an idea they meant any disrespect, it is quite probable that he may think it worth while

FIG. 138. JUPITER TERRACE.

to travel to New York and interview the writer. As our wayside friend explained to us, this guide always shot first and enquired afterwards. He had to go to court—not a lady, but a judge—and we failed to get him. We have no doubt, however, that court was properly administered, for we have not heard of any Montana judge being shot lately. After this interview, with the usual introduction and usual adieu, we rode off in the grey morning on our route to Mammoth Hot Springs.

We passed Cinnabar Mountain and the Devil's Slide. If his satanic majesty did slide here, he is extremely thin-skinned, and that is contrary to popular belief, as the face of this hill has a red streak where the blood flowed freely. Our wayside friend insisted we must stop at Gardiner City, and we agreed and rode on, looking for the City Hall, as the proper place to disembark, if one can use the term in regard to a horse. We fully expected to be received by the mayor, board of aldermen, &c. We found, however, they were very much like Mr. Gilbert's Pooh Bah, they were all concentrated into one individual, and so we compromised and agreed to be received by him; he was in his shirt sleeves to distinguish him from many others who had no shirts; and was the physician in ordinary, in other words he kept a drug store " in all that the name implies." We learned afterwards he had acquired the art of Esculapius from service in the New York police force, which is said to be the finest in the world. The stock was fastened to the counter for fear of accident, and might have been carried under one arm. The counter was " on trestles," to use an engineering phrase, and the floor was sanded, indeed, it was sanded by nature and well done; probably the sand extended some feet, at all events it was ankle deep. The doctor promptly carried us behind the drug store, and then we saw the uses of the drug business. Some one said the uses of adversity were sweet, but so are those of a drug store in Gardiner City. We were offered chairs (boxes) and requested to "shout." It being early and our throats full of cobwebs, we "shouted" and then naturally smoked and "made the rounds." It might be said that Gardiner City was entirely a city of tents, although its inhabitants were not like the Rechabites in any other respects. Indeed Gardiner City did not wish to be anything but tents just then, and the decision was a wise one, as the boundary line of the Yellowstone Park was at that time in process of settlement, and it was subsequently moved about three or four miles and Gardiner City was set back just that distance; not at all a difficult matter, but one that might have proved so had the town been more permanent in its character. A short ride brought the party to the Mammoth Springs, and here an illustration occurred of the confidence the natives have either in their ability to right themselves in case of betrayed trust, or else their great trust in human nature.

On the way one of the party dropped his rifle and broke the stock badly. When we came to Yankee Jim's, that worthy loaned our friend a fine heavy breechloading revolver, and when he was joked with the fact that he would never see his revolver again, he said he knew a man when he saw him.   Here again at Mammoth Springs the same trust was shown. We had been advised to leave all our money and valuables behind us, and the writer had but a few dollars and a draft on a bank.   Some man he was introduced to offered to cash this draft, which was nearly 100 dols., and did so.   Our wayside friend took us up the various terraces formed by the

FIG. 139.   A BASIN FORMATION.

overflowing of the hot water and the deposition of the soda held in solution until mound after mound has arisen.   In many cases the deposits form natural bath tubs of hot water.   The bath-houses were primitive, being made of rough plank, and the method of ascertaining if they were vacant was equally primitive ; you simply looked in, and if there was no occupant it was yours till you wanted to get out.   Our friend looked upon these adjuncts of pressing civilisation with scorn and said he could show us natural bath tubs far better.   So we climbed the third terrace, the white glistening soda being more trying to the eyes than snow, and finally found a natural formation just long enough for a man to lie down in.   The sensation was most delightful.   The water was quite hot enough to be soothing, and the natural tendency would have been to go to sleep, but there was only one tub there, and while the last man could have been granted this privilege, the others could not, and we passed a very pleasant hour in bathing and arranging matters of importance.   These springs are said to be highly beneficial for inflammatory rheumatism, the waters being saturated with various mineral substances, and our verdict was that this is very probably true, since various sorenesses and chafings due to over 66 miles riding seemed to leave us as by magic.   We all felt renewed, in a word were new men, and the new men were thirsty, so went to the nearest shanty and "shouted."   Then our wayside friend was kind enough to give us his opinion of us, and further to regret our inability to get a suitable guide, and finally to tell us that if we would go into the park by the Gardiner River route, it was quite possible he could arrange matters so as to go with us.   It is certain we decided this point instanter, and he then arranged that we should leave as soon as our "outfit" came up, and if we failed to secure him, he promised to find some one who would take us over Mount Washburn, and by waiting till our party came from the opposite direction, we could return with them.   The point of this is, that the Yellowstone Park is a rectangle 65 miles by 55—a good idea of it may be obtained by studying the map—and Mount Washburn, which is on one side rising 11,000 ft. above the sea level, is impassable except on horseback.   Our party, or to use the vernacular, "our outfit," having teams, could only go around two sides of the rectangle to the foot of Mount Washburn and then return by the same route, but we proposed to enter the park from the other side crossing Mount Washburn, and continuing on the route which had brought them to its foot.   It may be said here, that the Gardiner River route is one of the most beautiful and grand portions of this park.   The cañons are deep and the country is very much broken.   The views are superb, and the water abundant in every direction.   We took a farewell look at the Mammoth Springs, not forgetting the Liberty Cap, in which the heart of all true Americans should delight.   This curious formation was originally a geyser, and by constant eruptions and building up from sediment has risen to the height of 52 ft.   The resemblance at a distance to a liberty cap is really quite effective and distinct.   If there had only been a "spread eagle" somewhere around, patriotism would have overflowed like the waters of the Mammoth Springs.   Or if there had only been something that looked like an eagle, but there was nothing of the sort, the Bird of Freedom either thought it was too hot for him there, or else he was extinguished by the Liberty Cap.

At a little distance from the hotel is a most curious formation known as Hymen's Terrace (see Fig. 141). Although at first glance the figures seem to be walking on ice, yet the support under their feet is anything but cool. In fact, the water is quite hot, and it trickles along from the bubbling hot springs, and by depositing sediment it is gradually raising a formation

FIG. 140. THE CAP OF LIBERTY.

similar to the terraces at the Mammoth Springs. We left the Mammoth Springs about 2 P.M., and rode along the road to Clarke's Forks. We could plainly see in the distance on a side hill, the glitter of the falls of the Gardiner River, although ten miles distant. We numbered four, our guide and his friend, and we two engineers. We were all well mounted the writer having the only full-bred Indian pony, dubbed from his former owner, an old chief of the Crow Indians, "Old Crow." His ears were split at the ends, which is the mark of the Crow Indians. The others rode half-bred horses, and it was strange to note the difference even in blood. We passed over a prairie-dog town, whose little burrows were thickly put together, and being covered with long grass were perfect pitfalls for anything but an Indian pony. He invariably jumped over them and never stumbled, which was more than could be said for any other horse of the expedition. Suddenly our leader called a halt, and the horses stopped instantly. The guide jumped from his saddle, rifle in hand, and took a

FIG. 141. HYMEN'S TERRACE.

position very close to the head of Old Crow, and as he sighted along the barrel of his rifle, sitting as I did in the saddle, I looked along it also, but look as I might I saw nothing.   I braced myself for the report, not knowing what Old Crow would do, but beyond a switch of his tail he never budged, but a wild cat did, and gave a leap from a tree at a long distance and fell to the rocks.   These hunters seemed to have wonderful eyes, and could see anything and everything within a range which was incomprehensible to us " tender-feet."

After three hours' riding and climbing we came to Gardiner River Falls. This fall is 250 ft. high, making two leaps to the bottom of the chasm. Although small in width it is extremely beautiful, and the rain having been abundant, and the melting snows having swollen the river, it appeared at its best   We could have lingered a long time, but were hastened by the thought of the long ride still before us.   Up hill and down valley we rode, with walls rising on either side to great heights, till just when we were beginning to think even holidays had hard work, we saw considerably below us a valley which looked as peaceful and quiet as any poet's imagination could possibly dream of.   Near the foot of the slope where we were to descend was a log cabin, and from its stick chimney the smoke was curling. It certainly looked attractive after our hard ride, and the various battles we had been having with the mosquitoes on our journey hither.   Alas ! for the fallibility of human predictions, the mosquitoes in the lovely valley exceeded in numbers and ferocity any met in our trips.   It was possible they had been kept on short rations of late and saw their opportunity and meant to improve it.   A shout at the cabin brought out a tall man with a black beard sprinkled with grey ; his complexion was brown, and he looked very much like some of our enemies in the late war, to wit a " bush whacker." Strange to say he proved to be a Southerner and to have fought in their army, but that was rather a bond of union to the present writer, who had frequently seen his kind in 1862.   He was " Uncle John," and as such we welcomed him and his invitation to " light and hitch."   Our horses were stripped of saddle and bridle and promptly proceeded to roll over and over in the long grass.   A Montana proverb says : " A roll is as good as a feed of oats to a cayeuse."   Uncle John proved to be a good cook, and his bread rose at the proper time.   He always cooked his meat in the pan he washed his dishes in, which was a true application of the principle of the economy of labour, and we had some antelope for supper which we thoroughly appreciated.   Sleeping was difficult owing to the mosquitoes, that came in swarms into the house for the warm fire, but out of doors did not annoy us. Our horses had long lariat ropes attached and draped them while feeding on the rich grass.

At early dawn three of us started, leaving my companion to hunt with Uncle John till our return in two days.   We soon reached Barronett's Bridge, and the keeper being absent we opened the bridge ourselves and proceeded ; the Yellowstone River is extremely deep and swift at this point, and the rise on the other side up to Specimen Mountain

2 E

is very steep.   At about 11 A.M. we reached the East Fork of the Yellowstone River, which is quite broad and rapid ; there is no bridge here, and the river must be crossed in a skiff.   The horses were stripped of everything, including bridle and halter, and driven to the edge of the river by long poles, with which the entire party splashed in the water until the horses were off their feet.   It needed nothing more, for the stream was running at least ten miles an hour, and the horses struck out for the other shore with all possible effort, and landed about a quarter of a mile below their point of entry.   We crossed in a flat-bottomed skiff with saddles, &c. Our ferryman simply kept the boat broadside to the surging waters, and we were rapidly swept down the stream, landing near to the horses.   At the top of the bluff an old Missourian met us.   He was about 6 ft. 3 in. and bent somewhat with age.   He looked like the traditional guerilla of our war, and, strange to say, such he proved to be, having been a member of Quantrell's band, which was one of the worst in the war.   We were told he had been obliged to get out to the territories as soon as the war ended. However, one is not particular under the circumstances, and as the writer was to spend the day and night in " Uncle Billy's " cabin, he made up his mind to cultivate Uncle Billy.   We had with us the " universal persuader," warranted to suit every man in Montana, and Uncle Billy was not the man to throw good liquor over his shoulder, and, moreover, he was a kindly intentioned man and glad to see us.   The writer was left there for the day and night while the other two, the guide and his friend, carried out their original intention of riding to Clark's Forks, some thirty miles distant, then, if our guide so arranged matters, he was to return to Uncle Billy's and we were to go to Uncle John's and pick up my companion and make the trip.   If the guide could not return, the writer was to come back to Uncle John's alone, and he and his companion were to go on with him.   All the various points of demarcation were carefully impressed on the tourist that he should not fail to find his way back to Uncle John's, and yet it was with a deep sense of loneliness he saw the guide ride off, for though but a two days' acquaintance yet he seemed a link binding us to civilisation.   Being very tired, and ready to rest, a bed seemed the best place even at 11 A.M., but the mosquito thought it was also his opportunity, and sleep was impossible.   The only thing to do was to get into a draught and build a smudge fire, and sit in the smoke.

In the afternoon Uncle Billy, feeling that as host he must entertain his guest, proposed to go trout fishing, and also to visit Soda Butte.   The

writer lamented the absence of a fine trout-rod, but Uncle Billy said they
"had no use for it," as they drove the trout on to the grass. Not wishing
to display any ignorance, this was taken as the usual method of angling, but
it was food for thought all the way out, and even more food for thought on
the return.

Soda Butte is a mound some 30 ft. high, caused probably by the overflowing of a hot spring and constant deposit of soda. At the time it was

FIG. 142. OBSIDIAN CLIFFS.

inactive, but proved to be full of interest, in which respect it resembled
certain beds we had slept in. Near by was a wide running stream in which
was a beaver colony and dam. Going up this stream a mile—and it was
cold and beautifully clear—we came to a lake into which it emptied. A
small dam was hastily thrown up across the outlet, and the water speedily
spread over a large tract of ground, perhaps two or three acres. The two
men then went up the stream with long poles and beat down towards this
dam, and when they had splashed along the bed of the stream and stood
in it about knee-deep, still splashing with their poles, the dam was cut and

the water rapidly fell into the lake, while on the grass lay something over sixty beautiful trout gasping and ready to be cooked.

On our return to Uncle Billy's cabin, the writer complained bitterly of the mosquitoes, and recourse was again had to the smudge, but Uncle Billy solemnly stated that by 5 P.M. they would depart, and he proved a true prophet, for no sooner had the sun sunk behind the bluff opposite than these disturbers of peace vanished.

At midnight a trampling of horses' feet proclaimed the return of the guide, who had ridden sixty miles since 11 A.M. The next morning we took leave of Uncle Billy, who seemed like an old friend, and at the ford the ferryman refused to take my money either for himself or for the ferriage, saying that it was Uncle Billy's orders, and if he found that his instructions to this effect had been violated he would shoot him. On reaching Barronett's Bridge, we again let ourselves across, and had gone a short distance when we heard a hail and saw a man running towards us with a rifle in his hand. The guide said we had better stop, and we found he only wanted to collect the toll for crossing the bridge, and he charged for both ways, because he argued we must have crossed once to get on the side we were, as he knew we came from the springs. It is no use to argue with a man who carries a rifle unless you shoot first, so we paid and went on. The next day we all went hunting and obtained two elk. The meat of this animal is certainly the most delicious ever eaten. It has the flavour of venison, only is much more delicate and tender. On the following day we crossed Mount Washburn on a game trail at an elevation of 10,000 ft. above the sea level. In the pass at the summit was a snow-drift some 20 ft. deep, at the edge of which Old Crow baulked and would not venture till he had thoroughly satisfied himself the crust would bear. Although it was the fourth day of July, yet we enjoyed the novelty of a snowball fight, and thus celebrated the birthday of our country. Just before beginning the ascent of Mount Washburn we made a detour to visit Tower Falls. These falls are named from the peculiar buttes which extend along the cañon for a little distance above the fall. The water drops perpendicularly about 160 ft., and is almost like spray at the bottom. After descending Mount Washburn and riding for a long time we heard a distant rumble, which announced our approach to the Great Falls of the Yellowstone. The excitement was such we could feel our hearts beat loudly, and each one pressed onward at a rapid pace ; but the road required caution, for it was only a trail, but as the writer drew nearer to its roar it seemed as if nothing

FIG. 143. TOWER FALLS.

FIG. 144. THE UPPER FALLS.

could restrain his impatience, so, putting spurs to Old Crow, he rushed
ahead in the direction of the sound and caught a glimpse of the falls
through the wood.   Dismounting hastily, for it was no longer possible to
approach on horseback, these wonderful falls burst upon the view.   The
picture showing them (see Fig. 145) is taken at a distance of three-
fourths of a mile, and from the summit of a rock well-named Point Lookout.
The plunge is 364 ft., or about twice that of Niagara.   The photograph
dwarfs their appearance, which is simply overpowering ; the gazer is
absolutely stunned by the sight.   The foam and spray rise up at least
100 ft. from the bottom, and the wind occasionally lifts it and shows the
plunging water at the base.   It was possible afterwards to climb down the
face of the rocks to the very brink of the precipice and to look down the
sheet from the jutting rock squared off in the right-hand side of the picture.
An artist friend went some distance up the river and essayed to pole over
the stream on a raft.   When he reached the centre he found to his horror
that the pole would not touch bottom, and that he was slowly and surely
floating to the brink.   He told the writer that for a moment he seemed to
lose all consciousness, and then he revived, and made frantic efforts to reach
the shore ; just as the raft commenced to get into the rapids, which are
about half a mile up, and from which there would have been no escape, he
succeeded in getting a hold on the bottom again and finally in getting
ashore.   It made one shudder to contemplate the scene even in imagination.
The sides of the cañon are of varied colours, almost like the solar spectrum,
red being the predominant one.   Our guide urged us to climb, at once, up
the cañon, before night set in, that we might have all the next day for the
falls, here known as the Lower Falls, and those three-fourths of a mile
distant known as the Upper Falls.   The path was rough and stony, and
after skirting the edge of the cañon for over two miles we at last reached
a point from whence we had the extended view of the Grand Cañon of the
Yellowstone (see Figs. 146, 147, and 148).   At the point from which
this view is taken the cañon is 1500 ft. deep, and has a most gloomy
and overpowering majesty.   The river below seems a mere thread, although
it must be quite wide, and is very swift.

We reached our camp well tired out, and slept soundly in Uncle John's
buffalo robes.   This was on the night of July 4, and so great was the
elevation that the cold was sufficient to freeze stiff the lariat ropes which
held our horses.   The next morning I had a realising sense of the rarity of
the atmosphere, for after climbing down the bank by the camp and taking a

FIG. 145.   POINT LOOKOUT.

FIG. 146. GRAND CAÑON FROM POINT LOOKOUT.)

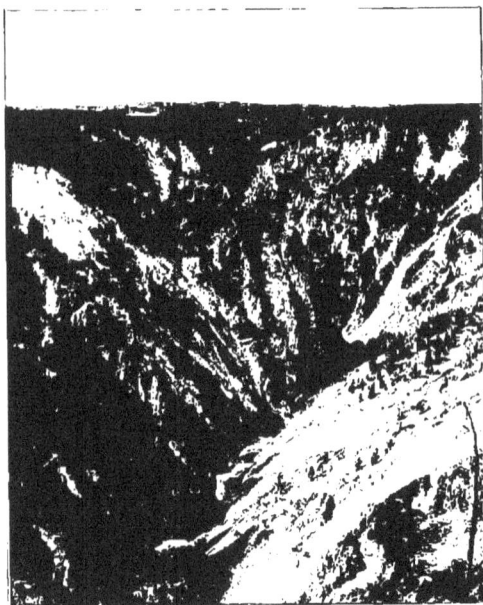

FIG. 147. THE GRAND CAÑON.

FIG. 148. GRAND CAÑON.

242

morning plunge, I essayed to climb up, feeling very much refreshed by the cold water. On scrambling to the top of these rocks I was so exhausted that I threw myself on the ground almost fainting, and was restored by the guide, who came quickly with the "invigorator," and said one must remember that at such an elevation climbing must be more gradually undertaken. After a delightful breakfast on elk, we proceeded to climb down to the brink of the falls, where we sat for hours ; but here we had to part with Uncle John, who had proved a good friend to us, and we really felt sad to see the kind-hearted old man go slowly over the hill, leading behind him the horse which had brought our camp utensils. We did the right thing by him, however, dividing carefully our last ration of "invigorator" with him, as we assured him we should meet our party somewhere that day *en route* for the falls, and as we had left them with five gallons we had no doubt of an abundant supply. With this explanation and assurance on our part, he consented to a stirrup cup, and so we parted, wishing him all speed and no end of good luck. About three-fourths of a mile distant we came upon the Upper Falls, some 260 ft. high, but much wider and far more furious than the Lower Falls. There is a majesty about the Lower Falls the Upper Falls do not possess, although they are very beautiful. Here, too, it is possible to climb to the very edge of the rock and look down on the sheet of water. On leaving these falls, which we did after a few hours, we commenced to descend the mountain rapidly, and shortly after starting came into a violent snowstorm, albeit it was July 5. As we further descended the snow changed to hail, which pelted us furiously, and even the ordinarily immovable Old Crow commenced to show signs of weariness. Our guide finally decided to halt for a few moments, as it was highly important we should save our horses, for we had quite a long journey before us, and having no camp equipage, or even blankets, must reach the lower geysers by night. While resting we ate what is known as jerked elk meat. This is apparently dry, and looks unpalatable, but after a little chewing it seems to acquire moisture, and really is quite enjoyable if one is hungry enough. Even had it not been so, we had no other choice, although we had some elk meat with us, for there was no shelter, and it would have been impossible to have cooked anything in the storm. Fortunately the hail ceased, giving place to a drizzling rain, which eventually cleared somewhat, when we saw below us Sulphur Mountain. Our guide promptly declared that the specks we saw moving along its base were our "outfit," or, in other words, the companions we had left at Mammoth Springs several days previously. He even pronounced that one man was riding a white horse, and the sequel proved he was quite correct, although only a good glass revealed to our shorter vision all he could see without any artificial aid. On meeting our friends a halt was made, and the experiences of each party exchanged. Their's was peculiar, novel, and interesting, and merits a description at some length. After we (my companion and myself) had left them at Mammoth Springs, they had endeavoured to obtain a guide, and had experienced some trouble, as a party of "tender-feet" had been around abusing the country, wearing two abominations, a plug hat and a boiled shirt, and demanding all sorts of things of the natives. They had been quieted, however, by a threat from some guide that he would bore them full of holes if they did not "light out," so they "lit." This episode had created a little prejudice against "tender-feet," and our friends were only able to obtain the services of a young fellow, who was no earthly good to them, and who promptly demanded the employment of a night herder, whose duties were to see the horses did not stray away, but who used to sleep soundly in his tent and let the cattle take their chances. In every party each man who is not useful is a terrible nuisance, and the guide and the night herder were no exception. They were always going to show the party bears, elk, &c. In one instance they were successful, they did show them a bear. The guide ran across him by accident, and he being one of the curious sort of bears, and of an inquiring disposition, ran after the guide, and followed him up to the camp, and then seeing the party, fled without giving them a shot. The guide was an ingenious sort of chap, and being rather mortified at his fright, undertook to explain it by saying it was simply a part of his plan to lure the bear to the camp, so that the hunters could get a shot at him. This did not increase the respect of the party or their teamsters for the courage of the guide, and the next day they had a further chance to lament his want of pluck. When the party left St. Paul, six small kegs each containing a gallon of Old Crow (not horse-flesh, but rye whisky) had been procured as a precaution against snake bites, not that we had seen any snakes, but "an ounce of precaution is worth a pound of cure," and so by analogy a gallon of Old Crow is worth, well, almost anything poisonous in the Yellowstone. When we left our friends at the Mammoth Springs, I had suggested we should take a keg on our saddle, but was overruled, and we only filled our flasks. When our friends reached the geysers, they went out to view these great wonders, and left the camp in charge of the teamsters. It was the fourth day of July, and that in America is "the day we celebrate," so the whole party, guide, teamsters, cook, night herder and all, proceeded to fill up. Finally, some old matter was recalled, which had occurred years before in some past time, between the ancestors of the cook and those of one of the teamsters, at least our friends believed it was some ancestral feud, because each one reviled the other's progenitors. At last the cook naturally flew to a knife and the teamster to a hatchet. Fortunately, each man was so drunk he could not hurt the other, and the fight degenerated into tearing each other's clothes; the cook seemed to be the most expert, for he stripped the teamster to one garment. The party of tourists were well armed, but seemed to forget this fact, and stood by in great anxiety. On hearing this story, the writer very naturally thought their anxiety arose from the danger each contestant stood in from his antagonist, but my informant told me he did not care much for that, as the death of either in itself would have been a small loss. "But," said he, "there was our cook, the man we depended on to feed us, in bodily peril. If he had been killed or disabled what would we have done for our living?" So human nature is selfish and always brings everything to a personal issue. The men were, however, readily separated, and when so parted showed at once their appreciation of the national motto, "Divided we fall," for they fell down and were ignominiously bundled into their tents, where they slept off their drunken fever, and were afterwards good friends all the rest of the trip. However, our friends decided not to put themselves in any further peril of this kind, and so instead of guarding against the evil by guarding the source of it, they took axes and smashed in the heads of the kegs, baptizing the sand in the geyser region with the spiritual comfort provided for man.

This tale was related to us at Sulphur Mountain and affected us as a new saddle does a sore-backed steed, for when we saw our friends, we had fondly anticipated the warmth of Old Crow, after our chilling ride of fifteen miles that morning. All was in vain; so we sadly went on our way, and as Bunyan says, "We saw them no more." For we were to go on to the geysers and they to the Great Falls, and as we continued our trip back to the Mammoth Springs while they had to double back from the falls, we should be several days ahead of them.

It now set in to rain very hard, and as the writer had providentially or otherwise lost his rubber coat the day before, so it bid fair to be a bad trip. We rode some 15 miles through a pitiless storm, even Old Crow (the horse this time and from this on) looking very much discouraged. The region was a gloomy one, being volcanic in its character, and what shrubbery there might have been was actually blasted as though by fire. We passed forests of trees all shrivelled and blackened, and groves of small trees leafless and bare. It looked extremely like some of Doré's illustrations in Dante. At last we saw way beyond us, probably five miles distant across a plain, a log house, and our tired horses were urged towards it. The writer reached it first, and with difficulty dismounted, he seeming to be almost part of the saddle, and a very stiff part too. Inside was a roaring fire in a huge stove, which took a log some 3 ft. long, and it was full and red hot. The proprietor afterwards explained that he had seen us with a glass when we had first struck the plain and knew we were "tender-feet," and expected us there. He looked to be a good-natured man, but was in no sense cordial and never expressed any interest in us or our plans. He proved subsequently to be most genial, and we were loth to leave ; his at first repellant manners were explained on the ground he had also been visited by the "dude" party, who had been so disagreeable at Mammoth Springs, and he thought we were some more of them. There was little in our party to suggest a dude. We wore the flannel shirt and felt hat of the country, and we were wet through and generally forlorn. We asked if he had any spirits, and explained the reason of our shortness of this important article. Then he seemed to wake up, but not on our account. "Great heavens !" said he, learning of the geyser episode and the loss of the liquor, "what a waste, what a waste ; six gallons you said, O Lord ! O Lord !" This did not, however, warm or cheer us, but the frontiersmen take time, and this was not our first experience with them, so we waited and shivered. Presently he brought in a stone jug corked (?) with a corn-cob and placed it on the table. The writer seized a tin cup and took a liberal amount. Having lived on the borders of civilisation at various times, he was not unfamiliar with all sorts of fluids of a strong nature. But this fellow was a giant. He felt as if he had swallowed a torch-light procession, for it not only burned all the way down but thereafter, and the tears flowed freely down his cheeks. It had the desired result, however, and started up the chilled blood. When we learned that the concoction was composed of brandy and gin in equal parts we did not wonder at its effect on the lining of the throat. It was here we enjoyed another refreshing hot bath from the water of a small boiling spring. It was arranged so that it could be turned into a tub, and there was a large barrel of cold water convenient to temper it with, for it was boiling hot. The next morning our host insisted in hitching up his double team and taking us around. He declared our horses needed rest, and that he wanted to "stand treat." He was a thorough man, and we liked him extremely. First he showed us the Paint Pots. These are mud geysers, and boil up mud of the most brilliant colours and varied hues. We then proceeded to the geysers, and their number may be conceived when in a square mile fifty are located, and the height to which they throw the streams of boiling water, is from 50 ft. to

FIG. 149. THE "PAINT POTS."

250 ft. The strangest feature of the whole is the fact that geysers adjacent to each other, have different intervals. The first one we saw was the Grand. This was not in operation just then, but was steaming, and might go off any time within the next few hours. Its stream is nearly 300 ft. high, and measures 6 ft. in diameter. We then passed the Grotto Geyser, which is nearly opposite, and where the formation by deposit has covered up the outlet, and afterwards came to the Splendid. Several men were here awaiting its eruption, and one of them told us he had thrown a pair

of heavy army trousers into the geyser, but they had sunk out of sight. This geyser was boiling furiously, and at intervals would throw up a spurt of water and steam about 10 ft. to 15 ft. It was evidently near its eruption. Each spurt came at a shorter interval, and finally there was a terrific roar, and the geyser burst into activity. The stream continued to keep at a maximum elevation for several minutes, the trousers rising to the top of the column, and looking very much like a truncated man, or rather a man sans trunk ; their owner uttered an exclamation of delight,

Fig. 150. The Grotto Geyser Cone ; Upper Basin.

and declared they had been well washed, and had no doubt they had gone to the antipodes. The Giant Geyser was also boiling full, but it did not go off till some hours afterwards ; this stream, although large and powerful, rises only to about 150 ft. All these geysers after eruption fall gradually till they become simply a boiling spring, and the water flows over their edge and streams away in a hot rivulet. Comet Geyser is also a very beautiful one, and is large, although not so high as some. Several pictures of these phenomena are given on the following pages. The formation

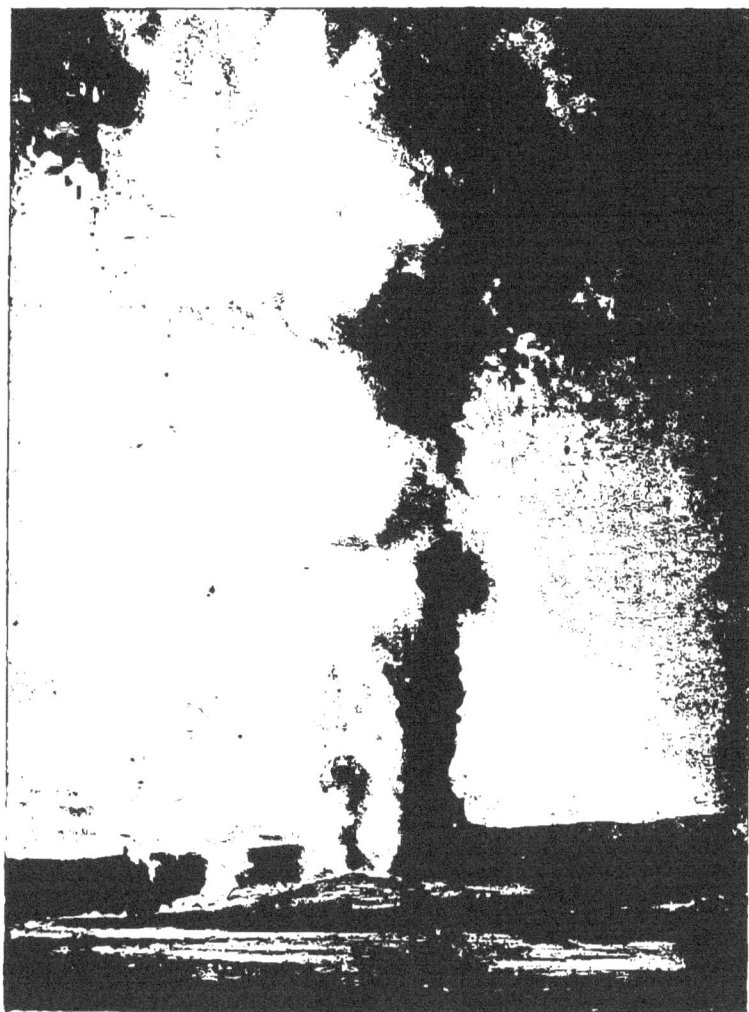

FIG. 151. THE GIANT GEYSER.

FIG. 152. CASTLE GEYSER.

FIG. 153. COMET GEYSER.

of Castle Geyser is peculiar, and is evidently caused by deposit of the sediment, which soon becomes hard. The writer picked up a soft piece, and before two days it had become quite hard, although at first it was almost like clay, and had to be dried on a piece of board to keep its shape. And now we come to one of the largest and most peculiar of the geysers, viz., "Old Faithful," so called from the fact that it has the most regular interval of any geyser. It comes into eruption at intervals of from 55 minutes to an hour. It stands on the summit of a small hill and can readily be seen even at a distance. The column thrown is 5 ft. in diameter and rises to a height of 200 ft. The water in the cone is remarkably clear and beautiful, and when it becomes quiet after an eruption, it is possible to look down 50 ft. into its depths, and the colour is of the most beautiful blue ever seen. Indeed, a painting the writer once saw of this cone was so brilliant that no one, unless it was Mr. Ruskin or the late Mr. Turner, would have believed it was anything but an exaggeration of nature. A short distance from Old Faithful is a small mound of an oval shape hence known as the Beehive ; near it is an opening perhaps a few inches in diameter, and distant from the cone about 10 ft. We found two men watching this small opening, which it seems, is an indicator ; when the indicator bubbles and spurts, the geyser will go off within half an hour, and the Beehive indicator was hard at work. As it is a remarkably beautiful geyser, and only goes off three times in twenty-four hours, we were anxious to see it, and our patience was rewarded, for it took its full half-hour, and then burst upwards to its full height of some 200 ft. Time fails to describe all the geysers we were fortunate enough to see, but enough has been shown to enable the reader to form a pretty correct idea of the wonders of this region. The next day we visited Fairy Falls, where the water falls over a precipice 250 ft., and comes down in a gentle rain. And now, as time pressed, the two travellers resolved to accomplish what even the old guides said was a feat, and an impossibility for any but a veteran to make any show of success in. We were at this particular time 110 miles from Livingstone, and we told our host we proposed to ride out in two days so as to intercept the train on the Northern Pacific Railroad which passes in the morning. This was only possible by reaching Livingstone the night previous. Our horses had several days' rest, and had been fed with oats ; there was not much doubt of their ability to do this, but when it came to the endurance of the rider, all the old settlers shook their heads and said no " tender-foot " could do it. For purposes of convenience in sleeping, as we

FIG. 154. BEEHIVE GEYSER.

FIG. 155. GROUP OF GEYSERS.

had no camp equipages, we were obliged to limit the first day's journey to 44 miles. We started, regretting to leave our kind host, and dined at the Falls of the Gibbon, a beautiful sheet of water about 50 ft. in height. We had no special adventure, although the journey was full of interest, and we saw many beautiful streams. We arrived safely at Gardiner City and slept in blankets and in the sawdust. It was soft and comfortable, indeed we could have slept on a pinnacle. The next morning we had to part with our faithful friend and guide, who had been our constant companion for two weeks. We had hard work to make him receive any money for his services, although this had been arranged before, and, finally, when all was fixed, he seemed to feel rather humiliated notwithstanding ; we assured him he had well earned it, and said, " Well, I ought to take it, I suppose, for I am in debt and need it." Next fall he sent the writer a magnificent elk's head with seven prongs, a Rocky Mountain sheep's head and skin, and a silver tip bear cub's skin, which serve to constantly recall his kindness and devotion. We reached Livingstone at 8.30 that night, having ridden 66 miles since 5.30, and having rested two hours at Fridley's Ranche to feed our horses and ourselves, and to purchase ordinary beer at 75 cents per bottle. Here the writer thought to indulge in a little political economy ; finding beer to be 25 cents per glass, and calculating a bottle would make four glasses at least, though they are very small, he decided it would be economy to buy a bottle. It proved to hold five glasses, and as neither he nor his companion cared for more than two, the cook, who stood looking in the door, was asked to empty the bottle. He promptly accepted the proposition so far as drinking was concerned, but said he did not like beer and would take whisky, if it was all the same to the gentlemen. It was " all the same " to them, although it demanded an additional quarter. On reaching Livingstone, although 8.30 P.M., it was light enough to read a letter while the writer sat in the saddle, for he felt that to dismount was never to get up again, and this letter demanded an immediate answer by telegraph ; so Old Crow's head was turned towards the telegraph station, at least a quarter of a mile distant, yet such is the endurance of these Indian ponies that he actually galloped over there and back as though he had not just put 66 miles of up and down hill, some of it very steep, behind him. After a good rest we took the train the next morning, and our Yellowstone trip became a thing of the past. In concluding, it must be said that the present facilities for this trip are very different from those described.

FIG. 156. "OLD FAITHFUL" GEYSER.

FIG. 157. THE FALLS OF THE GIBBON.

The railroad runs to the park by a branch from Livingstone, and the tourist can get everything he wants at the Mammoth Springs Hotel, including horses and outfit generally.   Moreover, there are several other hotels at convenient intervals through the park, and the roads are good. In other words, civilisation brings all these wonders within easy reach even of Europe and at a moderate price.   The North Pacific Railway will take an Englishman at Liverpool, give him five days in the park, and return him to Liverpool in twenty-eight days.   The writer would not advise anyone to take so short a time there, but very much may be accomplished in ten days at the park, especially now the facilities are so great for moving about.   It is worth remembering that the trout are good, and bears are to be had for the shooting.

There is probably no spot in the world offering so many wonderful sights within so small an area.

PRINTED AT THE BEDFORD PRESS, 20 AND 21, BEDFORDBURY, STRAND, LONDON, W.C.

www.ingramcontent.com/pod-product-compliance
Lightning Source LLC
Chambersburg PA
CBHW021525210326

41599CB00012B/1384